EERL Withdrawn
Surplus/Duplicate

The Invariant Imbedding Theory
of Nuclear Transport

THE ELMER E. RASMUSON LIBRARY
UNIVERSITY OF ALASKA

Modern Analytic *and* Computational Methods *in* Science *and* Mathematics

A GROUP OF MONOGRAPHS
AND ADVANCED TEXTBOOKS

Richard Bellman, EDITOR
University of Southern California

Published

1. R. E. Bellman, R. E. Kalaba, and Marcia C. Prestrud, Invariant Imbedding and Radiative Transfer in Slabs of Finite Thickness, 1963

2. R. E. Bellman, Harriet H. Kagiwada, R. E. Kalaba, and Marcia C. Prestrud, Invariant Imbedding and Time-Dependent Transport Processes, 1964

3. R. E. Bellman and R. E. Kalaba, Quasilinearization and Nonlinear Boundary-Value Problems, 1965

4. R. E. Bellman, R. E. Kalaba, and Jo Ann Lockett, Numerical Inversion of the Laplace Transform: Applications to Biology, Economics, Engineering, and Physics, 1966

5. S. G. Mikhlin and K. L. Smolitskiy, Approximate Methods for Solution of Differential and Integral Equations, 1967

6. R. N. Adams and E. D. Denman, Wave Propagation and Turbulent Media, 1966

7. R. L. Stratonovich, Conditional Markov Processes and Their Application to the Theory of Optimal Control, 1968

8. A. G. Ivakhnenko and V. G. Lapa, Cybernetics and Forecasting Techniques, 1967

9. G. A. Chebotarev, Analytical and Numerical Methods of Celestial Mechanics, 1967

10. S. F. Feshchenko, N. I. Shkil', and L. D. Nikolenko, Asymptopic Methods in the Theory of Linear Differential Equations, 1967

11. A. G. Butkovskiy, Distributed Control Systems, 1969

12. R. E. Larson, State Increment Dynamic Programming, 1968

13. J. Kowalik and M. R. Osborne, Methods for Unconstrained Optimization Problems, 1968

14. S. J. Yakowitz, Mathematics of Adaptive Control Processes, 1969

15. S. K. Srinivasan, Stochastic Theory and Cascade Processes, 1969

16. D. U. von Rosenberg, Methods for the Numerical Solution of Partial Differential Equations, 1969

17. R. B. Banerji, Theory of Problem Solving: An Approach to Artificial Intelligence, 1969

18. R. Lattès and J.-L. Lions, The Method of Quasi-Reversibility: Applications to Partial Differential Equations. Translated from the French edition and edited by Richard Bellman, 1969

19. D. G. B. Edelen, Nonlocal Variations and Local Invariance of Fields, 1969

20. J. R. Radbill and G. A. McCue, Quasilinearization and Nonlinear Problems in Fluid and Orbital Mechanics, 1970

21. W. Squire, Integration for Engineers and Scientists, 1970

22. T. Parthasarathy and T. E. S. Raghavan, Some Topics in Two-Person Games, 1971

23. T. Hacker, Flight Stability and Control, 1970

24. D. H. Jacobson and D. Q. Mayne, Differential Dynamic Programming, 1970

25. H. Mine and S. Osaki, Markovian Decision Processes, 1970

26. W. Sierpinski, 250 Problems in Elementary Number Theory, 1970

27. E. D. Denman, Coupled Modes in Plasmas, Elastic Media, and Parametric Amplifiers, 1970

28. F. H. Northover, Applied Diffraction Theory, 1971

29. G. A. Phillipson, Identification of Distributed Systems, 1971

30. D. H. Moore, Heaviside Operational Calculus: An Elementary Foundation, 1971

31. S. M. Roberts and J. S. Shipman, Two-Point Boundary Value Problems: Shooting Methods

32. V. F. Demyanov and A. M. Rubinov, Approximate Methods in Optimization Problems, 1971

33. S. K. Srinivasan and R. Vasudevan, Introduction to Random Differential Equations and Their Applications, 1971

34. C. J. Mode, Multitype Branching Processes: Theory and Applications, 1971

35. R. Tomovic and M. Vukobratovic, General Sensitivity Theory

36. J. G. Krzyz, Problems in Complex Variable Theory

37. W. T. Tutte, Introduction to the Theory of Matroids, 1971

38. B. W. Rust and W. R. Burrus, Mathematical Programming and the Numerical Solution of Linear Equations

39. J. O. Mingle, The Invariant Imbedding Theory of Nuclear Transport.

40. H. M. Lieberstein, Mathematical Physiology.

The Invariant Imbedding Theory

of Nuclear Transport

John O. Mingle

Kansas State University, Manhattan, Kansas

American Elsevier Publishing Company, Inc.

New York London Amsterdam

QC
721
M732

AMERICAN ELSEVIER PUBLISHING COMPANY, INC.
52 Vanderbilt Avenue, New York, N.Y. 10017

ELSEVIER PUBLISHING COMPANY
335 Jan Van Galenstraat, P.O. Box 211
Amsterdam, The Netherlands

International Standard Book Number 0-444-00123-9

Library of Congress Card Number 73-187687

© American Elsevier Publishing Co., Inc., 1973

All rights reserved.
No part of this publication may be reproduced,
stored in a retrieval system, or transmitted
in any form or by any means, electronic,
mechanical, photocopying, recording,
or otherwise, without permission in
writing from the publisher,
American Elsevier Publishing Company, Inc.,
52 Vanderbilt Avenue, New York, N.Y. 10017.

Library of Congress Cataloging in Publication Data

Mingle, John O
    The invariant imbedding theory of nuclear transport.

    (Modern analytic and computational methods in
science and mathematics, 39)
    Bibliography: p.
    1. Neutron transport theory. 2. Gamma rays--
Scattering. 3. Invariant imbedding. I. Title.
II. Series.
QC721.M732   539.7'213   73-187687
ISBN 0-444-00123-9

Manufactured in the United States of America

CONTENTS

PREFACE

In searching for a title for this book, I have attempted to signify the combined fields of neutron transport and gamma-ray transport by the phrase "nuclear transport." Since both of these fields can be considered a subset of a more general "particle transport," this book could have been entitled "The Invariant Imbedding Theory of Particle Transport." Because it is also written to introduce invariant imbedding to the reactor theory group and the shielding group of the nuclear engineering field, the word "nuclear" seemed more appropriate.

In order to enhance the communication that this book would have with the nuclear engineering field, I have chosen to use the nomenclature generally of neutron transport theory; however the analogous nomenclature is indicated for gamma-ray transport theory when appropriate. Naturally, there are other differences between neutron and gamma-ray transport theory in that they obey different physical laws. When a particular phenomenon is considered that is peculiar to only one field, such as the concept of neutron criticality theory, its limited application is at all times stressed.

Another purpose of my writing is to present sufficient results and details of invariant imbedding theory so that researchers in this field can use this book as a framework and general background reference. It is also hoped that they may find some interesting, unanswered questions concerning invariant imbedding theory in this book and be challenged to find the solutions. As an example, Chapter 11 merely introduces the application of invariant imbedding theory to curved geometries and many questions are still to be resolved.

At this point, I should like to express my appreciation to several of the many persons who have helped me either directly or indirectly with this writing. First is Dr. Richard E. Bellman who introduced me to the field of invariant imbedding and who convinced me to write this book. Next is Dr. William R. Kimel, who, as my department head in nuclear engineering at Kansas State University for many years, gave me much personal encouragement. A valuable member of my writing team has been my excellent typist, Mrs. Joan Hart, who, after many years, has mastered the mystery of my handwriting. Finally, my wife, Patricia, and my children, Elizabeth and Stephen, who have endured my preoccupation with this task, my sincerest thanks to all.

JOHN O. MINGLE
Manhattan, Kansas

CHAPTER 1

INTRODUCTION TO TRANSPORT THEORY

1.1. THE PHYSICAL PROCESS OF TRANSPORT THEORY

The general concept of "transport theory" refers to a physical process whereby a particle migrates through a lattice composed of scattering centers. The first important part of this concept is that the migrating material can be considered to be a particle such as a molecule, an electron, a neutron, or a photon. Of these four examples only the photon is by nature a nonparticle-like material, although when quantum mechanical considerations are employed, it acts like a particle. A neutron or molecule changes its speed of travel as its total energy changes, whereas a photon always moves with the speed of light but its frequency or wavelength changes as its total energy varies. In particular, high-energy photons in the x-ray and gamma-ray range are considered in nuclear transport processes.

The second part of the transport concept is that the material through which the particle moves interacts with the particle in a scattering-like process. When the migrating material can be considered a particle, the idea of a scattering collision is natural; however, only the net overall result of the interaction needs to "look-like" a scattering process. For instance, a neutron may suffer an elastic collision with an atomic nucleus and undergo an elastic scattering; on the other hand, it may be absorbed by the nucleus and then it or a like neutron be readmitted at a lower energy level in a different direction and thus suffer an inelastic scattering. From the overall view both of the processes appear to be a scattering phenomenon.

In order to include all transport processes it is necessary to generalize the concept of scattering to include absorption as a special case. Once absorption occurs then a law-of-readmission is necessary and this can range widely in its physical content. For instance, a neutron absorption by a nucleus can result in the emission of another neutron, one to several gamma rays, another mass particle like a proton or an alpha particle, or finally a fissioning of the nucleus into several new particles. Sometimes a combination of these events occur. As another example the absorption of a gamma photon by an atom can result in the emission of an electron, a positron, a neutron or another photon. The simultaneous emission of an electron and a positron is referred to as "pair-production" and occurs only when high-energy gamma photons are considered.

It is noted that in a scattering interaction the original particle or one like it reappears after the interaction, but its energy has been changed, usually to a lower total energy content, and it now has a new direction of travel. Whereas in an absorption process a different type of particle may appear but the original-type particle is lost.

When the transport of molecules is considered, the scattering processes are natural, except that the medium is not necessarily stationary, but the absorption phenomenon is replaced by a chemical reaction rate process. This physical process of molecular diffusion leads to the distinction between a "linear" and a "nonlinear" transport theory. In linear transport theory the equations are mathematically linear in the particle density of the migrating particles, whereas in nonlinear cases the equations must consider the probability of like-particle interactions and thus quadratic or higher terms appear. An example of the latter case is the molecule-molecule interactions of common molecular diffusion theory. In this book only the linear transport theory of nuclear processes will be utilized; thus interactions such as neutron-neutron will not be considered. The physical meaning of this linear transport theory is that the migrating particle does not significantly disturb the medium through which it is passing. These assumptions are particularly valid for the study of nuclear processes that are associated with the technology of neutron chain reactors.

1.2. THE CONCEPT OF CROSS SECTIONS

The mathematical description of a transport process depends upon parameters that describe the various types of interaction probabilities for the medium as well as a scattering law that predicts energy and angular changes for the migrating particle. The first concept is described by a "cross section" for the material. Let

σ = "effective target area" for an interaction with the migrating particle with the units cm^2/nucleus in the case of neutrons but cm^2/electron in the case of photons.

This σ is referred to as the "microscopic cross section" for a particular type of atom and is usually obtained by experiment. If the cross section is large, then the nucleus "looks-like" a large target area as far as the particle is concerned. In neutron transport theory the units of this microscopic cross section are usually given in terms of "barns" where one barn is 10^{-24} cm^2. In order to change from the consideration of a single nucleus a "macroscopic cross section" is introduced as

$$\Sigma = N\sigma, \qquad (1.2\text{-}1)$$

where N is the nucleus or electron density per unit volume for the material as a whole and for neutrons becomes

$$N = \rho N_0/A. \qquad (1.2\text{-}2)$$

Here ρ is the density of the material, A is the atomic weight of the target medium, and N_0 is Avogrodo's number giving the number of atoms in an atomic weight of the material. For electrons it is also necessary to multiply by the Z-number of the atom to take into account the number of electrons per atom. Thus the units of Σ become cm^{-1} in metric units; however, a better way to picture it is as cm^2/cm^3 or the effective target area per unit volume of material. It is noted that for a mixture of target atoms the macroscopic cross sections are additive.

With any transport process the quantity of physical importance is the interaction rate and this becomes

$$I = nv\Sigma, \qquad (1.2\text{-}3)$$

where n is the density of the target particles, such as neutrons/cm^3, v is the speed of the particles in cm/sec and Σ is the cross section. Thus the interaction rate is expressed as interactions/cm^3, sec. This concept of the macroscopic cross section will be employed in this book to represent the probability of an interaction in transversing a known distance by considering the following

$$\frac{\Delta I}{nv} = \Sigma \Delta z. \qquad (1.2\text{-}4)$$

Here ΔI is the rate of interaction per unit area occurring in moving a distance Δz. Thus $\Sigma \Delta z$ is the probability of an interaction per unit area per unit particle flux, nv.

In use the macroscopic cross section, usually shortened to just cross section, may be for scattering, Σ_s; absorption, Σ_a; fission, Σ_f; capture, Σ_c; or total, Σ. In general

$$\Sigma = \Sigma_s + \Sigma_a \qquad (1.2\text{-}5)$$

and

$$\Sigma_a = \Sigma_c + \Sigma_f. \qquad (1.2\text{-}6)$$

Usually a quantity is defined giving the mean number of secondaries per collision which reduces to the probability of scatter in a nonfissioning medium and is

$$c = \Sigma_s / \Sigma. \qquad (1.2\text{-}7)$$

Naturally the absorption probability is

$$1 - c = \Sigma_a / \Sigma. \qquad (1.2\text{-}8)$$

The further breakdown of cross sections can be accomplished easily since they are an additive quantity. For instance, elastic and inelastic scattering cross sections are common for neutrons.

The numerical value of the cross section besides being a function of the composition of the material is a strong function of the energy of the migrating particle and thus this dependence will usually be noted by the functional notation $\Sigma(E)$. Some authors prefer to utilize the reciprocal of the cross section which is known as the mean free path, and this will be utilized later in the book when it is convenient to do so.

1.3. THE SCATTERING LAW

The second part of the description of the physical processes necessary for transport theory is a knowledge of what happens after the interaction. It is common to express this in functional notation as

$f(E, \Omega, E', \Omega')dE \, d\Omega =$ the probability of changing from an energy E' and a direction Ω' into the energy range dE about E and the direction $d\Omega$ about Ω once the appropriate interaction occurs.

This general description can allow for many individual forms for a given physical process. In particular Chapter 4 on scattering matrix concepts presents the forms utilized in this book for this scattering law.

Readers unfamiliar with nuclear processes should refer to books like Kaplan (1962), Glasstone and Sesonke (1963) or Weinberg and Wigner (1958).

1.4. TYPES OF TRANSPORT THEORY

Based upon the above physical principles transport theory can be classified in a number of ways. Some of these have already been mentioned; for instance linear and nonlinear transport. A natural classification is by types of particle such as neutron, photon, molecular, or electron. The category of electron transport does not entirely fit the above physical scheme since an electron having a charge will not travel in a straight path between interactions; however, certain subcases of electron transport can be well approximated by these restrictions.

Another manner of classifying transport theory is by the form of the equations. The two forms, classical transport theory and invariant imbedding transport theory, are contrasted in detail in Chapter 2. This book is concerned with the latter formulation for transport theory.

1.5. HISTORY OF INVARIANT IMBEDDING THEORY

The invariant imbedding theory can be interpreted as two different conceptional approaches to transport theory. One of these is considered as a physical approach whereby a particle counting process is applied to the basic physical concepts. This procedure was first utilized by Ambarzumian (1943) and applied to some extent by Chandrasekhar (1950) in his classic work. The first work with invariant imbedding in its title was by Bellman and Kalaba (1956) where they combined the principles of invariance with the functional multistage processes of dynamic programming of Bellman (1957). During the next several years a number of basic papers appeared by Bellman and his coauthors showing the invariant imbedding formulation and application to neutron and other particle processes. Perhaps the summary paper is Bellman, Kalaba, and Wing (1960). The bibliography of Rosescu (1966) lists these extended works of Bellman as well as many other authors during this period.

In the next period a series of application papers appeared using the invariant imbedding method. These are characterized by Dodson and Mingle (1964), Mingle (1966, 1967), Timmons and Mingle (1967), Kaiser and Mingle (1967) and Shimizu and Mizuta (1966). The first book containing considerable numerical results was by Bellman, Kalaba, and Prestrud (1963). This current work is the next extension in particle counting invariant imbedding transport theory.

The second approach to invariant imbedding is a mathematical approach that proceeded in parallel with the particle counting method. This mathematical approach transforms the common classical transport equations to the invariant imbedding form by the usage of functional analysis. Wing (1961) shows this and proceeds in more detail in Bailey and Wing (1964, 1965). In addition books such as Bellman (1967) and Denman (1970) have sections based upon this mathematical analysis.

Although this book stresses the particle counting process of invariant imbedding because it gives a better physical picture of the theory, both procedures if applied correctly lead to the same equations.

1.6. SUMMARY OF THE BOOK

The remainder of this book falls into three sections. Chapters 2-7 give the basic theory of the invariant imbedding method for slab geometry in multienergy, multiangle formulations. Chapters 8 and 10 give special applications to monoenergetic theory and photon transport, respectively. Chapters 9 and 11 give extensions of the basic theory, the first to multiparticle theory where coupled neutron-photon problems are formulated and the second to the special considerations necessary for curved geometries. The appendix shows some numerical procedures used to obtain typical results.

CHAPTER 2

BASIC INVARIANT IMBEDDING CONCEPTS

2.1. INTRODUCTION

In this chapter the theory of invariant imbedding processes is given for the simplest case, i.e., the so-called rod model. In this geometric simplification the angular variations that are normally present in nuclear transport theory are missing; therefore, the relationships that lead to the invariant imbedding formulation can be stressed. A more rigorous mathematical treatment of this rod model is presented by Wing (1962); whereas this treatment will stress the physical aspects. The more realistic cases will be covered in later chapters.

2.2. BASIC DEFINITIONS

In this nonenergy, nonangular model the only variable is the position along the rod; therefore, define the following:

$\Sigma(z)dz$ = The probability of an interaction at z in traveling a distance dz.

$1 - \Sigma(z)dz$ = the probability of not having an interaction at z in traveling a distance dz.

$c_F(z)$ = the probability of forward scattering at z once an interaction occurs.

$c_B(z)$ = the probability of backward scattering at z once an interaction occurs.

These quantities give the basic properties of the medium necessary for a transport process to occur. It is noted that the sum of c_F and c_B will be unity only if no absorption occurs in the medium.

2.3. THE CLASSICAL METHOD

In order to accentuate the differences between the invariant imbedding method and the normal classical transport theory, this latter form will be considered first.

Define the following partial currents:

$j^+(z)$ = the number of particles per unit time at z moving in the plus-z direction.

$j^-(z)$ = the number of particles per unit time at z moving in the negative-z direction.

In this particular case the usage of flux and current is the same since no angular variables are present; however, since the distinctions are normally different, the consistent usage of the term current will be employed here. In Chapter 3 the differences between the two terms are indicated.

The medium is to consist of the rod length between $0 \leqslant z \leqslant x$ as shown in Fig. 1, so that a particle balance around any dz increment within the medium produces

$$j^+(z + dz) = [1 - \Sigma(z)dz] \, j^+(z) + \Sigma(z)dz \; c_F(z)j^+(z)$$
$$+ \Sigma(z)dz \; c_B(z)j^-(z + dz) \tag{2.3-1}$$

and

$$j^-(z) = [1 - \Sigma(z)dz)] \, j^-(z + dz) +$$
$$\Sigma(z)dz \; c_F(z)j^-(z + dz)$$
$$+ \Sigma(z)dz \; c_B(z)j^+(z). \tag{2.3-2}$$

In order to evaluate the functional quantities a Taylor's series expansion is made. For instance, for $j^+(z + dz)$ this becomes

$$j^+(z + dz) = j^+(z) + \frac{d}{dz} j^+(z)dz + \cdots, \tag{2.3-3}$$

with a like expansion for the j^- function. Utilizing these expansions in Eq. (2.3-1) and Eq. (2.3-2) produces

$$\frac{d}{dz} j^+(z) = \Sigma(z) \{ [c_F(z) - 1] j^+(z) + c_B(z) j^-(z) \} \qquad (2.3\text{-}4)$$

$$-\frac{d}{dz} j^-(z) = \Sigma(z) \{ [c_F(z) - 1] j^-(z) + c_B(z) j^+(z) \}, \qquad (2.3\text{-}5)$$

where naturally all terms of order $(dz)^2$ and higher have been dropped. It is noted that for simplicity these equations do not contain any internal source provisions. The boundary conditions for Eq. (2.3-4) and Eq. (2.3-5) are chosen as

$$j^+(0) = 0 \qquad (2.3\text{-}6)$$

$$j^-(x) = 1. \qquad (2.3\text{-}7)$$

These conditions will make it convenient to show the connection between the invariant imbedding method and the classical method in a later section.

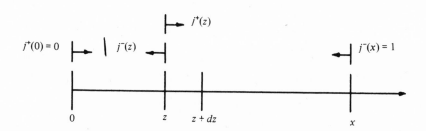

Fig. 1. Geometry of rod model for classical theory.

These classical transport equations besides representing the particle transport of either neutrons or gamma photons in a rod are also analogs of many electrical circuit equations; for instance, see Denman (1970).

It is noted that these classical equations are mathematically linear and of the boundary-value type. In order to solve this set of ordinary differential equations, the assumption is made that a homogeneous medium is being considered so that the total cross section and scattering probabilities are constants. Therefore, the system can be reduced to

$$\frac{d^2}{dz^2} j^+(z) + \Sigma^2 \left[c_B{}^2 - (c_F - 1)^2 \right] j^+(z) = 0,$$ (2.3-8)

the solution of which is

$$j^+(z) = A_1 \sin \kappa z + A_2 \cos \kappa z,$$ (2.3-9)

$$\kappa = \Sigma \sqrt{c_B{}^2 - (c_F - 1)^2}$$ (2.3-10)

and

$$j^-(z) = [\Sigma c_B]^{-1} \{ - [A_2 \kappa + A_1 \Sigma(c_F - 1)] \sin \kappa z$$
$$+ [A_1 \kappa - A_2 \Sigma(c_F - 1)] \cos \kappa z \},$$ (2.3-11)

where A_1 and A_2 are constants of integration. Naturally, if κ is imaginary, the hyperbolic functions are utilized. In the particular problem considered the boundary condition given by Eq. (2.3-6) makes A_2 zero and thus

$$j^+(z) = A_1 \sin \kappa z$$ (2.3-12)

$$j^-(z) = A_1 [\Sigma c_B]^{-1} [- \Sigma(c_F - 1) \sin \kappa z + \kappa \cos \kappa z].$$ (2.3-13)

Applying Eq. (2.3-7) gives

$$A_1 = [\Sigma c_B] [- \Sigma(c_F - 1) \sin \kappa x + \kappa \cos \kappa x]^{-1},$$ (2.3-14)

which completes the problem.

It is of interest to note that the two missing boundary values for this problem now become

$$j^-(0) = \kappa [- \Sigma(c_F - 1) \sin \kappa x + \kappa \cos \kappa x]^{-1}$$ (2.3-15)

$$j^+(x) = [\Sigma c_B] [- \Sigma(c_F - 1) \sin \kappa x + \kappa \cos \kappa x]^{-1} \sin \kappa x$$ (2.3-16)

There is a reduced case of this problem that has particularly simple expressions for answers; thus by letting $c_F = c_B = 1$,

$$j^-(0) = \sec \Sigma x \qquad (2.3\text{-}17)$$

$$j^+(x) = \tan \Sigma x. \qquad (2.3\text{-}18)$$

This represents the transmission and reflection respectively from a rod of length x when each interaction produces one particle moving in each direction.

Also from Eq. (2.3-16) the critical size for the rod becomes when $j^+(x)$ is infinite or

$$x = \kappa^{-1} \tan^{-1} \{\kappa [\Sigma(c_F - 1)]^{-1}\}. \qquad (2.3\text{-}19)$$

It is noted that this condition of a critical length is impossible in this problem if c_B is zero unless the rod is infinite in length.

2.4. THE INVARIANT IMBEDDING APPROACH

The invariant imbedding approach can be found in two equivalent ways. The first is mathematically by using the classical equations and converting them into the invariant imbedding form while the second is by deriving from physical principles the applicable result. In general, the physical derivation approach is preferred in this book, although the mathematical form is shown here for this simple model.

A convenient manner to show this relationship is the following: Let

$$j^+(z) = R(z)j^-(z). \qquad (2.4\text{-}1)$$

Substituting Eq. (2.4-1) into Eq. (2.3-4) and Eq. (2.3-5) produces the result

$$\frac{dR(z)}{dz} = 2\Sigma(z) [c_F(z) - 1]R(z) + \Sigma(z)c_B(z) [1 + R^2(z)]. \qquad (2.4\text{-}2)$$

Now since this holds for all z in the applicable range, let

$$z = x \qquad (2.4\text{-}3)$$

producing

$$\frac{dR(x)}{dx} = 2\Sigma(x)[c_F(x) - 1]R(x) + \Sigma(x)c_B(x)[1 + R^2(x)], \qquad (2.4\text{-}4)$$

where the boundary condition is

$$R(0) = 0 \qquad (2.4\text{-}5)$$

since Eq. (2.3-6) and Eq. (2.3-7) must hold.

It is noted that $R(x)$ as defined by Eq. (2.4-1) can be written as

$R(x)$ = the reflected particles per unit time from a rod of length x because of a unit input of particles to the rod at x.

Thus referring to Fig. 2 the five terms of order dx or less than can contribute to $R(x + dx)$ become:

Fig. 2. Geometry of rod model for invariant imbedding theory.

1. No interaction entering, reflection; no interaction leaving.
2. Interaction entering with backscatter.
3. Interaction entering with forward-scatter, reflection; no interaction leaving.
4. No interaction entering, reflection; interaction leaving with forward-scatter.

5. No interaction entering, reflection; interaction leaving with backscatter, reflection; no interaction leaving.

These contributions are shown in Fig. 3.

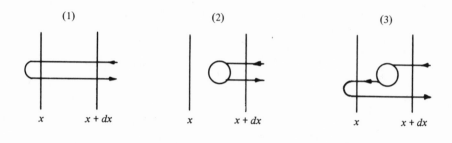

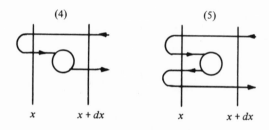

Fig. 3. The five interaction terms for the rod model
reflection function.

Utilizing the definitions given in Section 2.2 as applied to the distance dx produces

$$R(x + dx) = [1 - \Sigma(x)dx]R(x)[1 - \Sigma(x)dx] + \Sigma(x)dx\, c_B(x)$$

$$+ \Sigma(x)dx\, c_F(x)R(x)[1 - \Sigma(x)dx]$$

$$+ [1 - \Sigma(x)dx]R(x)\Sigma(x)dx\, c_F(x) \qquad (2.4\text{-}6)$$

$$+ [1 - \Sigma(x)dx]R(x)\Sigma(x)dx\, c_B(x)R(x)[1 - \Sigma(x)dx],$$

where the terms are in the same order as the above interaction list. Expanding in a Taylor's series and disregarding all terms of order higher than dx produces

$$\frac{dR(x)}{dx} = 2\Sigma(x)[c_F(x) - 1] R(x) + \Sigma(x)c_B(x)[1 + R^2(x)] \qquad (2.4\text{-}7)$$

with the natural boundary condition

$$R(0) = 0. \qquad (2.4\text{-}8)$$

This reflection equation is identical to the previous one, but has been obtained in an entirely different manner.

It is noted that the invariant imbedding equations are mathematically non-linear, but of an initial value type and thus a numerical solution immediately becomes feasible. Equation (2.4-7) is of the Riccati type having a quadratic term and in this simple unitary functional form a large amount of mathematical literature concerns it. For instance, see Bellman (1967).

Equation (2.4-7) has the analytical solution given by Eq. (2.3-16) since

$$R(x) = j^+(x) \qquad (2.4\text{-}9)$$

which can be verified by substitution. For more realistic transport models analytical solutions are not possible and numerical methods must be utilized.

A similar method can be utilized to produce the transmission result

$$\frac{dT(x)}{dx} = \Sigma(x)[c_F(x) - 1] T(x) + \Sigma(x)c_B(x)R(x)T(x) \qquad (2.4\text{-}10)$$

with

$$T(0) = 1 \qquad (2.4\text{-}11)$$

where the definition is:

$T(x)$ = the particles transmitted per unit time through a rod of length x because of a unit input at x.

Here naturally

$$T(x) = j^-(0) \qquad (2.4\text{-}12)$$

and thus Eq. (2.3-15) is the answer.

2.5. THE INTERNAL FUNCTIONS

Since it is possible to generate the invariant imbedding equations from the classical equations, it should be possible to reverse the procedure. By reference to Fig. 1 the following equations can be generated:

$$j^+(z) = R(z)j^-(z) \qquad (2.5\text{-}1)$$

$$j^-(z) = T(x - z) + R(x - z)j^+(z). \qquad (2.5\text{-}2)$$

The simultaneous solution of these equations is

$$j^+(z) = R(z)[1 - R(x - z)R(z)]^{-1} T(x - z) \qquad (2.5\text{-}3)$$

$$j^-(z) = [1 - R(x - z)R(z)]^{-1} T(x - z). \qquad (2.5\text{-}4)$$

Written in this manner the extension to the matrix case shown in later chapters is evident.

Similar expressions can be generated for the total current as

$$J(z) = j^+(z) - j^-(z) = [R(z) - 1] \ [1 - R(x - z)R(z)]^{-1} \ T(x - z) \qquad (2.5\text{-}5)$$

and the total flux as

$$\phi(z) = j^+(z) + j^-(z) = [R(z) + 1] \ [1 - R(x - z)R(z)]^{-1} \ T(x - z). \qquad (2.5\text{-}6)$$

2.6. EXERCISES

1. Generate the equivalent equations to Eq. (2.3-4) and Eq. (2.3-5) for the case where an internal source $S(z)$ is present. How is it necessary to define the source term?

2. Solve Eq. (2.3-4) and Eq. (2.3-5) for a homogeneous rod with the boundary conditions

$$j^+(0) = 1 \qquad (2.6\text{-}1)$$

$$j^-(x) = 0, \qquad (2.6\text{-}2)$$

and show what the applicable reflection and transmission functions become.

3. Denman (1970) sets up the full rod model as

$$j^-(0) = R_1(x) \, j^+(0) + T_1(x)j^-(x). \qquad (2.6\text{-}3)$$

$$j^+(x) = T_2(x)j^+(0) + R_2(x)j^-(x). \qquad (2.6\text{-}4)$$

Determine the equations that R_1, R_2 and T_1, T_2 satisfy when $j^+(0)$ and $j^-(x)$ are boundary conditions which are assumed nonzero.

4. Show the result for κ being imaginary in Eq. (2.3-19).

5. Derive Eq. (2.4-7) by the invariant imbedding physical interaction principals by writing the expression for $R(x + \Delta x)$ and taking the limit as $\Delta x \to 0$.

6. Show the figures similar to Fig. 3 for the interactions for the transmission function and write out the meaning of these terms.

7. Derive Eq. (2.4-10) by utilizing the relationship

$$j^-(x - z) = T(z)j^-(z). \qquad (2.6\text{-}5)$$

8. Derive Eq. (2.4-10) by using a unity input on the left end of the rod and still adding a distance dx to the right end.

9. Determine the internal partial currents for the generalized case shown in Exercise 3.

CHAPTER 3

ALBEDO THEORY: FUNCTIONAL FORM

3.1. INTRODUCTION

In order to derive the energy angle-dependent albedo for slab geometry, the usual particle counting technique is employed. The restriction to slab geometry is made only to accentuate the energy-dependent part so as not to mask the invariant imbedding principles by considering curved geometries.

3.2. BASIC DEFINITIONS

The following quantities are defined:

$\Sigma(x,E)dy$ = the probability of an interaction of any type occurring in traveling a distance dy at x for a particle of energy E.

$c(x,E)$ = the probability of a scattering occurring during an interaction at x for a particle of energy E.

$f(x,E,\Omega,E',\Omega')dEd\Omega$ = the probability of scattering into the energy range dE about E and into the solid angle $d\Omega$ about Ω from the energy E' and the direction Ω' at the position x.

$R(x,E,\Omega,E_0,\Omega_0)dEd\Omega$ = the reflected particle current* at x in the energy range dE about E and the direction $d\Omega$ about Ω because of a *unit* current input at the energy E_0 in the direction Ω_0 at x.

Current refers to the particles per unit area per unit time with the area measured on the surface of the medium. *Flux* refers to the same situation but the area is measured normal to the considered direction. In general a cosine multiplying factor exists between the two quantities.

This definition for the reflection function implies that Ω is outwardly directed while Ω_0 is inwardly directed, so that the functional would be identically zero for any other combinations. It is noted that this unity normalization is not that usually employed in radiative transfer, i.e., see Bellman (1963). These basic definitions allow the functional form of the particle albedo to be determined.

3.3 FIVE CONTRIBUTING TERMS

Consider the possible interactions occurring in dx of order dx or less as per Fig. 4:

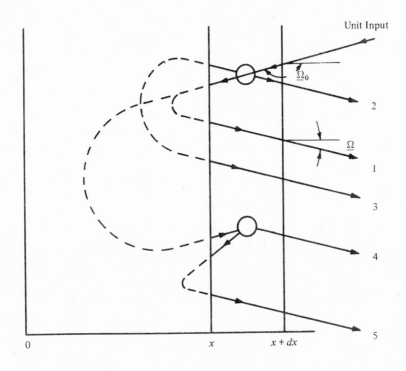

Fig. 4. The reflection function interactions.

1. No interaction entering, reflection, no interaction leaving.
2. Interaction entering with scattering outward.
3. Interaction entering with scattering inward, reflection, no interaction leaving.
4. No interaction entering, reflection, interaction leaving with scattering outward.
5. No interaction entering, reflection, interaction leaving with scattering inward, reflection, no interaction leaving.

The concept of entering and leaving is with respect to the distance dx. This leads to the following terms written in functional form.*

1. $[1 - \Sigma(x,E_0)dx/|\mu_0|] \, R(x,E,\Omega,E_0,\Omega_0)dEd\Omega \, [1 - \Sigma(x,E)dx/\mu]$,

2. $\Sigma(x,E_0)dx/|\mu_0|c(x,E_0)f(x,E,\Omega,E_0,\Omega_0)dEd\Omega$,

3. $\int_0^\infty \int_0^{4\pi} \Sigma(x,E_0)dx/|\mu_0|c(x,E_0)f(x,E',\Omega',E_0,\Omega_0)dE'd\Omega'R(x,E,\Omega,E',\Omega')$
$\cdot \, dEd\Omega[1 - \Sigma(x,E)dx/\mu]$,

4. $\int_0^\infty \int_0^{4\pi} [1 - \Sigma(x,E_0)dx/|\mu_0|] \, R(x,E'',\Omega'',E_0,\Omega_0)dE''d\Omega''$
$\cdot \, \Sigma(x,E'')dx/\mu'' c(x,E'')f(x,E,\Omega,E'',\Omega'')dEd\Omega$,

5. $\int_0^\infty \int_0^\infty \int_0^{4\pi} \int_0^{4\pi} [1 - \Sigma(x,E_0)dx/|\mu_0|] \, R(x,E'',\Omega'',E_0,\Omega_0)dE''d\Omega''$
$\cdot \, \Sigma(x,E'')dx/\mu'' c(x,E'')f(x,E',\Omega',E'',\Omega'')dE'd\Omega'R(x,E,\Omega,E',\Omega')$
$\cdot \, dEd\Omega \, [1 - \Sigma(x,E)dx/\mu]$.

Here $\mu = \cos\theta$ as per the customary notation†. Therefore,

$$R(x + dx,E, \Omega,E_0,\Omega_0)dEd\Omega = \text{sum of five terms.}$$

The net result when a Taylor's series expansion is utilized and terms of order higher than dx have been dropped so that the common term $dxdEd\Omega$ can be canceled is

---

*The FORTRAN notation of multiplication predominating is used throughout; thus s/av means $(s/a)v$.
†It is noted that the output direction μ is always positive while the inlet direction μ_0 is always negative and thus the absolute value signs.

$$\frac{d}{dx} R(x,E,\Omega,E_0,\Omega_0) = -[\Sigma(x,E_0)/|\mu_0| + \Sigma(x,E)/\mu] R(x,E,\Omega,E_0,\Omega_0)$$

$$+ \Sigma(x,E_0)c(x,E_0)f(x,E,\Omega,E_0,\Omega_0)/|\mu_0|$$

$$+ \Sigma(x,E_0)c(x,E)/|\mu_0| \int_0^\infty dE' \int_0^{4\pi} d\Omega' f(x,E',\Omega',E_0,\Omega_0)$$

$$\cdot R(x,E,\Omega,E',\Omega')$$

$$+ \int_0^\infty dE'' \Sigma(x,E'')c(x,E'') \int_0^{4\pi} d\Omega'' R(x,E'',\Omega'',E_0,\Omega_0)$$

$$\cdot f(x,E,\Omega,E'',\Omega'')/\mu''$$

$$+ \int_0^\infty dE'' \Sigma(x,\ E'')c(x,E'') \int_0^\infty dE' \int_0^{4\pi} d\Omega'' \int_0^{4\pi} d\Omega'$$

$$\cdot R(x,E'',\Omega'',E_0,\Omega_0)$$

$$\cdot f(x,E',\Omega',E'',\Omega'') R(x,E,\Omega,E',\Omega')/\mu''. \qquad (3.3\text{-}1)$$

The initial condition is

$$R(0,E,\Omega,E_0,\Omega_0) = 0 \qquad (3.3\text{-}2)$$

as no reflection occurs for a zero-sized medium.

In the normally encountered situation the scattering function is of the reduced form

$$f(x,E,\Omega,E',\Omega')dEd\Omega = f(x,E,E',\Omega \cdot \Omega')dEd\mu \qquad (3.3\text{-}3)$$

so that the quantity of interest is the scattering angle, θ_s, and not the two individual directions. In this case the ϕ-dependence of the solid angle can be removed by letting

$$R(x,E,\mu,E_0,\mu_0)dEd\mu \ = \ \frac{1}{2\pi} \int_0^{2\pi} d\phi_0 \int_0^{2\pi} d\phi R(x,E,\Omega,\Omega_0)dEd\mu, \qquad (3.3\text{-}4)$$

where naturally $d\Omega = d\phi d\mu$. This assumes that the usage of a ϕ-independent input is also satisfactory. If this is not the case, then the full angle dependence must be employed. Equation (3.3-1) reduces to

$$\frac{d}{dx} R(x,E,\mu,E_0,\mu_0) = -[\Sigma(x,E_0)/|\mu_0| + \Sigma(x,E)/\mu] R(x,E,\mu,E_0,\mu_0)$$

$$+ \Sigma(x,E_0)c(x,E_0)f(x,E,E_0,\Omega \cdot \Omega_0)/|\mu_0|$$

$$+ \Sigma(x,E_0)c(x,E_0)/|\mu_0| \int_E^{E_0} dE' \int_{-1}^0 d\mu' f(x,E',E_0,\Omega' \cdot \Omega_0)$$
$$\cdot R(x,E,\mu,E',\mu')$$

$$+ \int_E^{E_0} dE'' \, \Sigma(x,E'')c(x,E'')$$
$$\cdot \int_0^1 d\mu'' R(x,E'',\mu'',E_0,\mu_0)f(x,E,E'',\Omega \cdot \Omega'')/\mu''$$

$$+ \int_E^{E_0} dE'' \Sigma(x,E'')c(x,E'')$$
$$\cdot \int_0^1 d\mu'' R(x,E'',\mu'',E_0,\mu_0)/\mu'' \int_E^{E''} dE' \int_{-1}^0 d\mu'$$
$$\cdot f(x,E',E'',\Omega' \cdot \Omega'') R(x,E,\mu,E',\mu')), \tag{3.3-5}$$

where the restriction to only down-scattering in energy has been imposed by the limits used in the energy integrations.

By this definition of the albedo, the total output for a given input is

$$R(x,E_0,\mu_0) = \int_0^{E_0} dE \int_0^1 d\mu \, R(x,E,\mu,E_0,\mu_0) \tag{3.3-6}$$

while the response to a distributed input surface source is

$$R(x,E,\mu) = \int_E^{E_0} dE_0 \int_{-1}^0 d\mu_0 S(x,E_0\,\mu_0) \, R(x,E,\mu,E_0,\mu_0). \tag{3.3-7}$$

3.4. SPECIAL CASES

The most important special case is that for a semiinfinite medium whereby there is no change produced by adding a thickness dx. Therefore,

$$\frac{dR}{dx} = 0 \tag{3.4-1}$$

and the right side of Eq. (3.3-1) or Eq. (3.3-5) is set equal to zero. This produces an integral equation to solve for the quantity

$$R(\infty,E,\mu,E_0,\mu_0) = R(E,\mu,E_0,\mu_0). \tag{3.4-2}$$

In actual calculations involving other quantities, such as the transmission, this semiinfinite result is utilized for low energies as the medium generally appears infinite to low energy particles before it does to higher energy particles. In a later chapter the discretized version of this solution will be discussed further.

Another special case occurs when an energy independent situation is utilized. This monoenergetic result for a particle of energy E_0 can be obtained from the general albedo by using

$$R(x,\mu,\mu_0)d\mu = \int_0^\infty dE\, R(x,E,\mu,E_0,\mu_0)d\mu\, \delta(E-E_0) \qquad (3.4\text{-}3)$$

and the energy independent scattering functional, $f(x,\mu,\mu')$. The result when this is applied to Eq. (3.3-5) is

$$\frac{1}{\Sigma(x)}\frac{d}{dx} R(x,\mu,\mu_0) = -[1/\mu + 1/|\mu_0|]\, R(x,\mu,\mu_0)$$

$$+ c(x)f(x,\mu,\mu_0)/|\mu_0|$$
$$+ c(x)/|\mu_0| \int_{-1}^0 d\mu' f(x,\mu',\mu_0)\, R(x,\mu,\mu')$$
$$+ c(x) \int_0^1 d\mu''\, R(x,\mu'',\mu_0)f(x,\mu,\mu'')/\mu''$$
$$+ c(x) \int_0^1 d\mu''\, R(x,\mu'',\mu_0)/\mu''$$
$$\cdot \int_{-1}^0 d\mu' f(x,\mu',\mu'')\, R(x,\mu,\mu'). \qquad (3.4\text{-}4)$$

Considerable simplification occurs by writing this albedo in terms of the mean free path

$$\tilde{x} = \int_0^x \Sigma(x)dx \qquad (3.4\text{-}5)$$

as well as the restriction to isotropic scattering

$$f(\mu,\mu_0)d\mu = d\mu/2. \qquad (3.4\text{-}6)$$

These produce the symmetric form

$$\frac{d}{d\tilde{x}}\hat{R}(\tilde{x},\mu,\mu_0) = -[1/\mu + 1/\mu_0]\,\hat{R}(\tilde{x},\mu,\mu_0) + \frac{c}{2}\,\Gamma(\mu)\Gamma(\mu_0), \qquad (3.4\text{-}7)$$

where

$$\Gamma(s) = 1 + \int_0^1 d\mu'\,\hat{R}(\tilde{x},s,\mu')/\mu' = 1 + \int_0^1 d\mu''\,\hat{R}(\tilde{x},\mu'',s)/\mu'' \qquad (3.4\text{-}8)$$

and

$$\hat{R}(\tilde{x},\mu,\mu_0) = \mu_0\, R(\tilde{x},\mu,\mu_0) \qquad (3.4\text{-}9)$$

and the input has been written with respect to the inward normal so that μ and μ_0 are both positive. Much of the application of invariant imbedding theory to date has been with this monoenergetic, isotropic scattering, restricted form of the albedo functional and some special cases will be considered in a later chapter.

3.5. EXERCISES

1. Derive Eq. (3.3-1) by starting with the basic definitions in the form of differences, i.e., $R(x + \Delta x, E, \Omega, E_0, \Omega_0)\Delta E\Delta\Omega$, and take the limit as $\Delta x \to 0$.

2. Using Eqs. (3.3-1) and (3.3-4) show the details in obtaining Eq. (3.3-5).

3. Show that the Chandrasekhar (1950) H-function for a semiinfinite medium for a monoenergetic system is defined by

$$R(\mu,\mu_0) = \frac{\mu\mu_0}{\mu + \mu_0} H(\mu)H(\mu_0) \qquad (3.5\text{-}1)$$

and determine an integral equation for it. What restrictions on the scattering functions are necessary?

4. What restrictions are put on the scattering function by the transformation used in going from Eq. (3.3-5) to Eq. (3.4-4)?

CHAPTER 4

SCATTERING MATRIX CONCEPTS

4.1. INTRODUCTION

In order to discuss the general theory of energy-dependent particle transport, the concept of a scattering matrix will be required. This represents a discretization of the functional form of a general scattering cross section and includes angle as well as energy dependence.

4.2. FUNCTIONAL FORMS

To define a scattering distribution function it is necessary first to recall the definition of the c-value as

$c(r)$ = the probability in an interaction at position r that a scattering interaction will occur, i.e., Σ_s/Σ_t.

In general this will include all types of scattering, i.e., elastic and inelastic; however, specific separate probabilities can be utilized. For instance, with gamma-ray transport a pair production interaction could be considered a type of scattering and an appropriate probability utilized.

The distribution function for scattering is then

$f(r,E,\Omega,E',\Omega')dEd\Omega$ = the probability of scattering into the energy range dE about E and into the solid angle $d\Omega$ about Ω from the energy E' and the direction Ω' at position r.

The normalization of the distribution function can then be

$$\int_0^\infty dE \int_0^{4\pi} d\Omega \, f(r,E,\Omega,E',\Omega') = 1 \qquad (4.2\text{-}1)$$

which must be true for all E' and Ω'. Naturally, a relationship can exist between E and Ω so that large portions of the functional space for this distribution function will be zero. This occurs, for instance, in the consideration of Compton scattering for photons.

An example of typical distribution functions for homogeneous media would include isotropic scattering as

$$f(r,E,\Omega,E',\Omega')\,dEd\Omega = f(E,E')\,dE\,(d\Omega/4\pi) \qquad (4.2\text{-}2)$$

while heavy atom elastic and isotropic neutron scattering is

$$f(r,E,\Omega,E',\Omega')\,dEd\Omega = \delta\,(E - E')dE\,(d\Omega/4\pi). \qquad (4.2\text{-}3)$$

Pair production "scattering" of photons can be considered by the distribution function

$$f(r,E,\Omega,E',\Omega')\,dEd\Omega = \delta\,(E - E_p)dE\,[1 + \delta(\Omega + \Omega_p)]\,d\Omega/4\pi, \qquad (4.2\text{-}4)$$

where E_p = 0.511 Mev, the annihilation photon energy, and $\Omega_p = -\Omega$ always. Here the assumption of isotropic emittance of the annihilation photons is used as well as negligible positron travel.

4.3. MATRIX FORMS

In order to put the scattering distribution function into a discrete form, the energy range is divided into a number of groups as per Fig. 5.

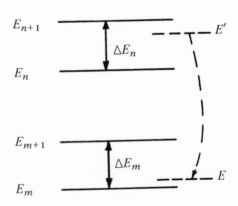

Fig. 5. Discrete energy group representation.

Therefore, define the following:

$$f_{mn}(r,\Omega,\Omega')d\Omega = \frac{1}{\Delta E_n} \int_{E_n}^{E_{n+1}} dE' \int_{E_m}^{E_{m+1}} dE f(r,E,\Omega,E',\Omega')d\Omega. \qquad (4.3\text{-}1)$$

Note that the integration over dE follows naturally, but the integration over dE' requires the division by ΔE_n in order to retain the proper normalization. This multigroup formulation implies that the sum over energy groups utilizing a first order numerical integration formula will result in the correct approximation to the integral over energy.

The discrete form for the angular dependence can proceed in a number of ways. The two most common are to expand the dependence in a few term series of spherical harmonics or to utilize a numerical integration formula. In general both procedures depend upon the assumption that the quantity of interest is only the scattering angle and not the independent directions Ω and Ω'. Thus let

$$\cos \theta_s = \Omega \cdot \Omega' \qquad (4.3\text{-}2)$$

so that

$$f_{mn}(r,\Omega,\Omega')d\Omega = \frac{1}{2\pi} f_{mn}(r,\cos \theta_s)d(\cos \theta_s), \qquad (4.3\text{-}3)$$

where the $1/(2\pi)$ is the Jacobian between $d\Omega$ and $d(\cos \theta_s)$.

In the first method the Legendre polynomial expansion is utilized as

$$f_{mn}(r, \cos \theta_s) = \sum_{l=0}^{L} \left(\frac{2l+1}{2}\right) f_{mn}^l(r)P_l(\cos \theta_s). \qquad (4.3\text{-}4)$$

For instance for linearly anisotropic scattering, the result is

$$f_{mn}(r, \cos \theta_s) = \frac{1}{2} f_{mn}^0(r) + \frac{3}{2} f_{mn}^1(r) \cos \theta_s \qquad (4.3\text{-}5)$$

since

$$f_{mn}^0(r) = \int_{-1}^{+1} d (\cos \theta_s) f_{mn}(r,\cos \theta_s) \qquad (4.3\text{-}6)$$

and

$$f_{mn}^1(r) = \int_{-1}^{+1} d(\cos \theta_s) \cos \theta_s \, f_{mn}(r,\cos \theta_s). \qquad (4.3\text{-}7)$$

If this is now applied to a system that is ϕ-independent, the utilization of the addition formula, see Weinberg and Wigner (1958),

$$\cos\theta_s = \cos\theta\,\cos\theta' + \sin\theta\,\sin\theta'\,\cos(\phi - \phi') \qquad (4.3\text{-}8)$$

produces the result

$$\int_0^{2\pi}d\phi\,\sin\theta\,d\theta f_{mn}(r,\cos\theta_s) = \int_0^{2\pi}d\phi\,\sum_{l=0}^{1}\frac{2l+1}{2}\,f_{mn}^l(r)P_l(\cos\theta_s)\sin\theta\,d\theta$$

$$= [\pi f_{mn}^0(r)\,\cos\theta\,\cos\theta']\,\sin\theta\,d\theta. \qquad (4.3\text{-}9)$$

Higher-order expansions ($L > 1$) do not generally yield such simple results. At this point the total scattering from one energy group to another for this model is

$$\int_0^{\pi}\sin\theta\,d\theta\,[\pi f_{mn}^0(r) + 3\pi f_{mn}^1(r)\,\cos\theta\,\cos\theta'] = 2\pi f_{mn}^0(r) \qquad (4.3\text{-}10)$$

which shows why the $1/(2\pi)$ is utilized in Eq. (4.3-3). Thus

$$\frac{1}{2\pi}\int_0^{4\pi}d\Omega f_{mn}(r,\Omega\cdot\Omega') = \int_{-1}^{+1}d(\cos\theta_s)f_{mn}(r,\cos\theta_s) = f_{mn}^0(r). \qquad (4.3\text{-}11)$$

Therefore, the quantity $f_{mn}^l(r)$ can be considered to be the basic quantity to be utilized. The scattering matrix can then be calculated in any desirable form from these coefficients and knowledge of the spherical harmonics.

For the alternate procedure the distribution function for the energy groups is discretized by a numerical integration formula. Thus if

$$f_{mn}(r,\Omega_i,\Omega_j) = f_{mn}^{ij}(r) \qquad (4.3\text{-}12)$$

then

$$\int_0^{4\pi}d\Omega f_{mn}(r,\Omega,\Omega') = \sum_i w_i\,f_{mn}^{ij}(r). \qquad (4.3\text{-}13)$$

where w_i is the appropriate weight for the position Ω_i.

Now in the important case where ϕ-independence occurs, this integration can be performed by considering Eq. (4.3-8). Thus solving for ϕ

$$\phi = \phi' + \cos^{-1}\left\{\frac{\cos\theta_s - \cos\theta\,\cos\theta'}{\sin\theta\,\sin\theta'}\right\} \qquad (4.3\text{-}14)$$

and

$$d\phi = -d(\cos\theta_s)\,\{1 - \cos^2\theta - \cos^2\theta' - \cos^2\theta_s + 2\cos\theta\cos\theta'\cos\theta_s\}^{-1/2}, \qquad (4.3\text{-}15)$$

so that
$$\int_0^{2\pi} d\phi f_{mn}(r,\cos\theta_s) =$$

$$2\int_{\cos\theta\cos\theta'-\sin\theta\sin\theta'}^{\cos\theta\cos\theta'+\sin\theta\sin\theta'} d(\cos\theta_s) \frac{f_{mn}(r,\cos\theta_s)}{[1-\cos^2\theta-\cos^2\theta'-\cos^2\theta_s+2\cos\theta\cos\theta'\cos\theta_s]^{1/2}} \quad (4.3\text{-}16)$$

This form is particularly convenient to use when the scattering angle is limited by the energy change as with Compton scattering of photons; for instance, when a δ-function involving $\cos\theta_s$ appears.

If the scattering data are recorded independent of ϕ then let

$$f_{mn}(r,\theta,\theta')d(\cos\theta) = \frac{1}{2\pi}\int_0^{2\pi} d\phi f_{mn}(r,\cos\theta_s)d(\cos\theta) \quad (4.3\text{-}17)$$

so that

$$\int_{-1}^{+1} d(\cos\theta)f_{mn}(r,\theta,\theta') = \sum_i w_i\, f_{mn}^{ij}(r), \quad (4.3\text{-}18)$$

where

$$f_{mn}^{ij}(r) = f_{mn}(r,\mu_i,\mu_j) \quad (4.3\text{-}19)$$

since $\mu = \cos\theta$. Therefore, the terms of the scattering matrix are the f_{mn}^{ij} for each homogeneous volume.

One special case occurs that should be noted. When either one of the discrete angles is zero, the addition formula, Eq. (4.3-8), reduces to

$$\cos\theta_s = \mu, \quad (4.3\text{-}20)$$

where θ' has been chosen zero. Therefore, the ϕ-integration is simply

$$\int_0^{2\pi} d\phi f_{mn}(r,\cos\theta_s) = 2\pi f_{mn}(r,\mu) \quad (4.3\text{-}21)$$

for this restricted case. Naturally, a similar expression exists for the case where $\theta = 0$. When $\theta = \theta' = 0$, then the result is

$$\int_0^{2\pi} d\phi f_{mn}(r,\cos\theta_s) = 2\pi f_{mn}(r,\pm 1), \quad (4.3\text{-}22)$$

where the plus sign is for forward scatter and the minus sign for backward scatter.

It is noted that for this discrete form a $\hat{f}$ will be employed to indicate backscattering while f will indicate forward scattering.

4.4. EXERCISES

1. In the general case where

$$f(r,E_m,\mu_i,\phi_k,E_n,\mu_j,\phi_l) = f_{mn}^{ikjl}(r)$$

deduce the discrete properties of these scattering matrix elements.

2. In the case where the general full spherical harmonics expansion is utilized, determine an expression for the coefficients. Note: Define the spherical harmonics functions in the orthonormal form.

ALBEDO THEORY: DISCRETE FORM

5.1. INTRODUCTION

Utilizing the scattering matrix concepts of the previous chapter, the discrete form of the reflection function or albedo is presented. This form is normally applied when numerical results via digital computer calculations are desired. Certain calculational aspects will be also considered in this chapter.

5.2. DISCRETE REFLECTION FORMULATION

In order to discretize the energy, angle variables, the method of numerical quadrature is employed as per the second form of the scattering matrix theory. Thus let

$$R(x,E_m,\mu_i,E_n,\mu_j) = R_{mn}^{ij}(x) \tag{5.2-1}$$

so that

$$\int_0^1 d\mu\, R(x,E,\mu,E_0,\mu_0) = \sum_{l=1}^L w_l R(x,E,\mu_l,E_0,\mu_0). \tag{5.2-2}$$

The quadrature weights and positions are usually taken as standard Gaussian integration results for the range $0 \leqslant \mu \leqslant 1$ although not limited to this by any means. Table 1 gives two abbreviated quadrature sets; the first a standard Gaussian and the second a Radau integration.

The energy discretization process generally proceeds by assuming certain energy groups, not necessary equally spaced, so that the midpoint formula or the Gaussian formula of order unity for a given group is employed. Thus utilizing backward difference notation

$$\int_E^{E_0} dE'\, R(x,E',\mu,E_0,\mu_0) = \sum_{k=1}^N \Delta E_k R(x,E_k,\mu,E_0,\mu_0), \tag{5.2-3}$$

where

$$\Delta E_k = E_k - E_{k-1} \tag{5.2-4}$$

and

$$\sum_{k=1}^{N} \Delta E_k = E_0 - E. \qquad (5.2\text{-}5)$$

In order to keep the treatment as general as possible the full ϕ-dependence will be included in the discrete formation and removed in a later section. Thus

$$f(x,E_m,\mu_i,\phi_k,E_n,\mu_j,\phi_l) = f_{mn}^{ikjl}(x) \qquad (5.2\text{-}6)$$

and

$$R(x,E_m,\mu_i,\phi_k,E_n,\mu_j,\phi_l) = R_{mn}^{ikjl}(x), \qquad (5.2\text{-}7)$$

and Eq. (3.3-1) takes the discrete form

$$\frac{d}{dx}R_{mn}^{ikjl}(x) = -[\Sigma_n(x)/\mu_j + \Sigma_m(x)/\mu_i]\,R_{mn}^{ikjl}(x) + \Sigma_n(x)c_n(x)\hat{f}_{mn}^{ikjl}(x)/\mu_j$$

$$+ \Sigma_n(x)c_n(x)/\mu_j \sum_{p=1}^{N} \Delta E_p \sum_{s=1}^{L} w_s \sum_{t=1}^{\hat{L}} \hat{w}_t\, f_{pn}^{stjl}(x)R_{mp}^{ikst}(x)$$

$$+ \sum_{q=1}^{N} \Delta E_q \Sigma_q(x)c_q(x) \sum_{u=1}^{L} w_u \sum_{v=1}^{\hat{L}} \hat{w}_v R_{qn}^{uvjl}(x)f_{mq}^{ikuv}(x)/\mu_u$$

$$\qquad\qquad\qquad\qquad\qquad\qquad\qquad\qquad\qquad (5.2\text{-}8)$$

$$+ \sum_{q=1}^{N} \Delta E_q \Sigma_q(x)c_q(x) \sum_{p=1}^{N} \Delta E_p \sum_{u=1}^{L} w_u \sum_{v=1}^{\hat{L}} \hat{w}_v \sum_{s=1}^{L} w_s \sum_{t=1}^{\hat{L}} \hat{w}_t$$

$$\cdot\, R_{qn}^{uvjl}(x)\hat{f}_{pq}^{stuv}(x)R_{mp}^{ikst}(x)/\mu_u \qquad (i,j = 1,2, \ldots, L;$$

$$k,l = 1,2,\ldots,\hat{L};\; m,n, = 1,2,\ldots N).$$

Here the $\hat{w}$ are the weights for the ϕ-integration over the range $0 \leqslant \phi \leqslant 2\pi$ with $\hat{L}$ being the order utilized. Also the inlet and outlet angles have been taken with respect to the input and output normal directions, respectively, thus keeping μ and μ_0 both positive in the range zero to one.

The reduction of Eq. (5.2-8) to the case of ϕ-independence can proceed either by discretizing Eq. (3.3-5) directly or by defining

$$R_{mn}^{ij}(x) = \sum_{k=1}^{\hat{L}} \hat{w}_k \sum_{l=1}^{\hat{L}} \hat{w}_l\, R_{mn}^{ikjl}(x) \qquad (5.2\text{-}9)$$

with a similar term for the scattering function and applying it to Eq. (5.2-8). The result is

TABLE 1

Numerical Integration Weights and Abscissas
for the Range Zero to Unity

Order	Gaussian[a]		Radau[b]	
	Position	Weight	Position	Weight
2	0.2113248654	0.5000000000	0.3333333333	0.7500000000
	0.7886751346	0.5000000000	1.0000000000	0.2500000000
3	0.1127016654	0.2777777778	0.1550510257	0.3764030627
	0.5000000000	0.4444444444	0.6449489743	0.5124858262
	0.8872983346	0.2777777778	1.0000000000	0.1111111111
4	0.0694138442	0.1739274226	0.0885879595	0.2204622112
	0.3300094782	0.3260725774	0.4094668644	0.3881934688
	0.6699905218	0.3260725774	0.7876594618	0.3288443200
	0.9305681558	0.1739274226	1.0000000000	0.0625000000
5	0.0469100770	0.1184634425	0.0571041961	0.1437135608
	0.2307653449	0.2393143353	0.2768430136	0.2813560151
	0.5000000000	0.2844444444	0.5835904324	0.3118265230
	0.7692346551	0.2393143553	0.8602401357	0.2231039011
	0.9530899230	0.1184634425	1.0000000000	0.0400000000

[a]Abramowitz et al. (1964).
[b]Calculated values.

$$\frac{d}{dx} R_{mn}^{ij}(x) = - \left[\Sigma_m(x)/\mu_i + \Sigma_n(x)/\mu_j \right] R_{mn}^{ij}(x)$$

$$+ \Sigma_n(x) c_n(x) \hat{f}_{mn}^{ij}(x)/\mu_j$$

$$+ \Sigma_n(x) c_n(x)/\mu_j \sum_{p=1}^{N} \Delta E_p \sum_{s=1}^{L} w_s f_{pn}^{sj}(x) R_{mp}^{is}(x)$$

$$+ \sum_{q=1}^{N} \Delta E_q \Sigma_q(x) c_q(x) \sum_{u=1}^{L} w_u R_{qn}^{uj}(x) f_{mq}^{iu}(x)/\mu_u \qquad (5.2\text{-}10)$$

$$+ \sum_{q=1}^{N} \Delta E_q \Sigma_q(x) c_q(x) \sum_{p=1}^{N} \Delta E_p \sum_{u=1}^{L} w_u \sum_{s=1}^{L} w_s R_{qn}^{uj}(x)$$

$$\hat{f}_{pq}^{su}(x) R_{mp}^{is}(x)/\mu_u \qquad (i,j = 1,2,...,L; \; m,n = 1,2,...,N).$$

The order of this system is now $L^2 N^2$ and thus still can be quite large depending upon the accuracy required.

5.3. SINGLE SCATTERING REMOVAL

In order to utilize as low an order of angular quadrature as possible, the expected angular-dependence must be a smooth slowly varying curve. This can be realized only if the first scattering result is removed from the equation by an analytical treatment. Thus noting the order-of-scattering expansion

$$R(x,E,\mu,E_0,\mu_0) = \sum_{\lambda=0}^{\infty} [c(x,E_0)]^{\lambda} R_{\lambda}(x,E,\mu,E_0,\mu_0) \qquad (5.3\text{-}1)$$

so that the reflection can be represented by the expansion

$$R(x,E,\mu,E_0,\mu_0) = R_1(x,E,\mu,E_0,\mu_0) + R_M(x,E,\mu,E_0,\mu_0). \qquad (5.3\text{-}2)$$

Here R_1 is the single scattered result and R_M is the multiscattered result. Utilizing this in Eq. (3.3-5) gives the coupled set

$$\frac{d}{dx} R_1(x,E,\mu,E_0,\mu_0) = -[\Sigma(x,E_0)/|\mu_0| + \Sigma(x,E)/\mu] R_1(x,E,\mu,E_0,\mu_0)$$

$$+ \Sigma(x,E_0) c(x,E_0) \, f(x,E,E_0,\Omega \cdot \Omega_0)/|\mu_0|, \qquad (5.3\text{-}3)$$

$$\frac{d}{dx} R_M(x,E,\mu,E_0,\mu_0) = -[\Sigma(x,E_0)/|\mu_0| + \Sigma(x,E)/\mu] R_M(x,E,\mu,E_0,\mu_0)$$

$$+ \Sigma(x,E_0) c(x,E_0)/|\mu_0| \int_E^{E_0} dE' \int_{-1}^{0} d\mu' f(x,E',E_0,\Omega' \cdot \Omega_0)$$

$$\cdot [R_1(x,E,\mu,E',\mu') + R_M(x,E,\mu,E',\mu')]$$

$$+ \int_E^{E_0} dE'' \Sigma(x,E'') c(x,E'') \int_0^1 d\mu'' [R_1(x,E'',\mu'',E_0,\mu_0)$$

$$+ R_M(x,E'',\mu'',E,\mu)] f(x,E,E'',\Omega \cdot \Omega'')/\mu''$$

$$+ \int_E^{E_0} dE'' \Sigma(x,E'') c(x,E'') \int_0^1 d\mu'' [R_1(x,E'',\mu'',E_0,\mu_0)$$

$$+ R_M(x,E'',\mu'',E_0,\mu_0)]/\mu''$$

$$\cdot \int_E^{E''} dE' \int_{-1}^0 d\mu' f(x,E',E'',\Omega' \cdot \Omega'') [R_1(x,E,\mu,E',\mu')$$

$$+ R_M(x,E,\mu,E',\mu')], \qquad (5.3\text{-}4)$$

where naturally

$$R_1(0,E,\mu,E_0,\mu_0) = 0 \quad \text{and} \quad R_M(0,E,\mu,E_0,\mu_0) = 0. \qquad (5.3\text{-}5)$$

The solution for R_1 is

$$R_1(x,E,\mu,E_0,\mu_0) = \exp\left[- \int_0^x dx' \tau(x',E,\mu,E_0,\mu_0)\right]$$

$$\cdot \int_0^x dx'' \left\{ \exp\left[\int_0^{x''} dx' \tau(x',E,\mu,E_0,\mu_0)\right]\right.$$

$$\left.\cdot \Sigma(x'',E_0) c(x'',E_0) f(x'',E,E_0,\Omega \cdot \Omega_0)/|\mu_0|\right\}, \qquad (5.3\text{-}6)$$

where

$$\tau(x,E,\mu,E_0,\mu_0) = \Sigma(x,E_0)/|\mu_0| + \Sigma(x,E)/\mu. \qquad (5.3\text{-}7)$$

In the general case of position-dependent properties this is usually solved numerically; however, in the case of layered homogeneous material

$$R_1(x,E,\mu,E_0,\mu_0) = R_1(a,E,\mu,E_0,\mu_0) \exp[\tau(E,\mu,E_0,\mu_0) \{a-x\}]$$

$$+ \Sigma(E_0) c(E_0) f(E,E_0,\Omega \cdot \Omega_0)/|\mu_0|[\tau(E,\mu,E_0,\mu_0)]^{-1}$$

$$\cdot [1 - \exp[\tau(E,\mu,E_0,\mu_0) \{a-x\}]]. \qquad (5.3\text{-}8)$$

Here the interested layer is between $a \leqslant x \leqslant b$.

In the more general case the discretized version of Eqs. (5.3-3,4) become

$$\frac{d}{dx} R_{mn1}^{ij}(x) = -[\Sigma_n(x)/\mu_j + \Sigma_m(x)/\mu_i] R_{mn1}^{ij} +$$

$$+ \Sigma_n(x) c_n(x) \hat{f}_{mn}^{ij}(x)/\mu_j, \qquad (5.3\text{-}9)$$

$$\frac{d}{dx} R^{ij}_{mnM}(x) = -[\Sigma_n(x)/\mu_j + \Sigma_m(x)/\mu_i] R^{ij}_{mnM}(x)$$

$$+ \Sigma_n(x)c_n(x)/\mu_j \sum_{p=1}^{N} \Delta E_p \sum_{s=1}^{L} w_s f^{sj}_{pn}(x) [R^{is}_{mp1}(x) + R^{is}_{mpM}(x)]$$

$$+ \sum_{q=1}^{N} \Delta E_q \Sigma_q(x)c_q(x) \sum_{u=1}^{L} w_u [R^{uj}_{qn1}(x) + R^{uj}_{qnM}(x)] f^{iu}_{mq}(x)/\mu$$

$$+ \sum_{q=1}^{N} \Delta E_q \Sigma_q(x)c_q(x) \sum_{p=1}^{q} \Delta E_p \sum_{u=1}^{L} w_u \sum_{s=1}^{L} w_s [R^{uj}_{qn1}(x) + R^{uj}_{qnM}(x)]$$

$$\cdot \hat{f}^{su}_{pq}(x) [R^{is}_{mp1}(x) + R^{is}_{mpM}(x)]/\mu_u , \tag{5.3-10}$$

whereas the layered homogeneous slab case result is

$$R^{ij}_{mn1}(x) = R^{ij}_{mn1}(a) \exp [\tau^{ij}_{mn}(a-x)]$$

$$+ \Sigma_n c_n \hat{f}^{ij}_{mn}/\mu_j [\tau^{ij}_{mn}]^{-1} [1 - \exp [\tau^{ij}_{mn}(a-x)]] \tag{5.3-11}$$

and

$$\tau^{ij}_{mn} = \Sigma_m/\mu_i + \Sigma_n/\mu_j . \tag{5.3-12}$$

The tradeoff in the size of the system in the general spatial-dependent case is interesting. For instance if $N = 5$, $L = 4$ is adequate for solving Eqs. (5.3-9,10) but $N = 5$, $L = 6$ is necessary for Eq. (5.2-10), then the total system size is 800 vs 900, respectively, and the single scattering approach still represents an advantage. However, the most advantage lies when Eq. (5.3-11) can be utilized so that the system size is 400 vs 900, a marked savings.

It is customary to define a modified reflection function that combines the energy delta increment into the albedo function definition. Thus let

$$\mathcal{R}^{ij}_{mn}(x) = R^{ij}_{mn}(x) \Delta E_m \tag{5.3-13}$$

and

$$\mathcal{f}^{ij}_{mn}(x) = f^{ij}_{mn}(x) \Delta E_m . \tag{5.3-14}$$

This produces the set of equations

$$\frac{d}{dx} R^{ij}_{mn1}(x) = -[\Sigma_n(x)/\mu_j + \Sigma_m(x)/\mu_i] R^{ij}_{mn1}(x)$$

$$+ \Sigma_n(x)c_n(x) \hat{f}^{ij}_{mn}(x)/\mu_j, \tag{5.3-15}$$

$$\frac{d}{dx} R^{ij}_{mnM}(x) = -[\Sigma_n(x)/\mu_j + \Sigma_m(x)/\mu_i] R^{ij}_{mnM}(x)$$

$$+ \Sigma_n(x)c_n(x)/\mu_j \sum_{p=1}^{N} \sum_{s=1}^{L} w_s f^{sj}_{pn}(x) R^{is}_{mp}(x)$$

$$+ \sum_{q=1}^{N} \Sigma_q(x)c_q(x) \sum_{u=1}^{L} w_u R^{uj}_{qn}(x) f^{iu}_{mq}(x)/\mu_u$$

$$+ \sum_{q=1}^{N} \Sigma_q(x)c_q(x) \sum_{p=1}^{q} \sum_{u=1}^{L} w_u \sum_{s=1}^{L} w_s$$

$$\cdot R^{uj}_{qn}(x) \hat{f}^{su}_{pq}(x) R^{is}_{mp}(x)/\mu_u, \tag{5.3-16}$$

where

$$R^{ij}_{mn}(x) = R^{ij}_{mn1}(x) + R^{ij}_{mnM}(x). \tag{5.3-17}$$

The similar result for Eq. (5.3-11) is obvious.

5.4. EXTENSIONS OF THE ALBEDO THEORY

One obvious extension to the albedo theory presented so far, which is generally applicable to neutron transport, is the case where two different particle types are represented with both types of interactions having a reasonable probability for a particular particle energy. The most straightforward case occurs when gamma-ray transport is considered and the energy is high enough (>1.02 MeV) so that both normal scattering and pair-production with subsequent annihilations are occurring. It is generally desirable to consider the pair-productions' interaction separately and predict a corresponding albedo. Naturally, normal reflection both before and after pair-production must be considered so that the new system of equations is coupled to the old one for regular scattering. This situation will be considered in a later chapter devoted to the special requirements of photon transport.

5.5. COMPUTATIONAL CONSIDERATIONS

Certain computational problems are considered in the appendix as they are generally applicable to the solution of many of the discrete formulations of this book.

5.6. EXERCISES

1. Formulate the theory equivalent to Eqs. (5.3-15, 16) for the discrete energy approach where a numerical quadrature formula, i.e.,

$$\int_E^{E_0} dE \; g(E) = \sum_{k=1}^{N} w_k g(E_k) \qquad (5.6\text{-}1)$$

is utilized for the energy integrations.

2. Formulate the theory equivalent to Eqs. (5.3-15, 16) for the discrete approach when it is desired to use a dimensionless energy-type variable as

$$\lambda = E_p/E, \qquad (5.6\text{-}2)$$

where E_p is a constant.

3. Utilizing Eqs. (5.3-15, 16) determine an expression for the "critical condition" of a two neutron group reactor with isotropic scattering assumed. Note: The c-value for neutrons can be represented by

$$c_n = \frac{(\nu\Sigma_f)_n \, \delta_{n1} + (\Sigma_s)_n}{\Sigma_n}, \qquad (5.6\text{-}3)$$

where ν is the neutrons per fission and Σ_f is the fission cross section. The δ_{n1} implies that fission neutrons are isotropically distributed in the first energy group only.

CHAPTER 6

TRANSMISSION THEORY

6.1. INTRODUCTION

In this section the transmission theory of particle penetration through a media is considered. The functional form of the transmission function is derived as well as the discrete form utilized for calculations.

6.2. BASIC DEFINITIONS

The transmission function is defined as:

$T(x, E, \Omega, E_0, \Omega_0)dEd\Omega^*$ = the transmitted particle current[†] emerging on the left slab face in the energy range dE about E and the direction $d\Omega$ about Ω because of a *unit* input current on the right face at the energy E_0 in the direction Ω_0 for a slab of size x.

It is noted that the input is at the right surface of the slab and that the dx increment will also be added at this right side. The output is on the left surface of the slab. The other basic quantities defined in Chapter 3 for the reflection function will also be utilized.

It is possible to define a conceptually different transmission function whereby the output is on the same side of the slab as the dx-increment; however, this produces the same functional equation as the defined transmission function so it will not be further considered here.

6.3. THREE CONTRIBUTING TERMS

In analogy with the reflection function derivation, the three transmission

---

*It is customary to define the solid angle at a surface so that outward angles from the surface have positive μ-values while inward directions have negative μ-values. Thus for this case although both solid angles are directed physically the same, because of this convention the input values have negative μ-values while the output ones are positive.

[†]See Section 3.2 for definition of current.

terms are:

1. No interaction entering, transmission.

2. Interaction entering with scatter inward, transmission.

3. No interaction entering, reflection, interaction leaving with scattering inward, transmission.

Figure 6 shows these various contributions. The resulting functional forms are:

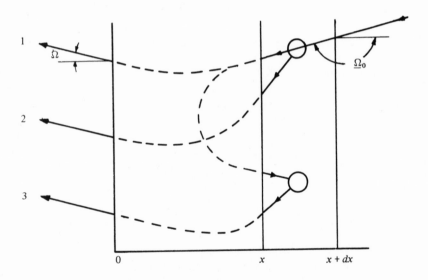

Fig. 6. The transmission function interactions.

1. $[1-\Sigma(x,E_0)dx/|\mu_0|]\ T(x,E,\Omega,E_0,\Omega_0)dEd\Omega$

2. $\int_0^\infty \int_0^{4\pi} \Sigma(x,E_0)dx/|\mu_0|c(x,E_0)f(x,E',\Omega',E_0,\Omega_0)\ dE'd\Omega'\ T(x,E,\Omega,E',\Omega')dEd\Omega$

3. $\int_0^\infty \int_0^\infty \int_0^{4\pi} \int_0^{4\pi} [1-\Sigma(x,E_0)dx/|\mu_0|]\ R(x,E'',\Omega'',E_0,\Omega_0)dE''d\Omega''$

    $\cdot\ \Sigma(x,E'')dx/\mu''c(x,E'')f(x,E',\Omega',E'',\Omega'')dE'd\Omega'$

    $\cdot\ T(x,E,\Omega,E',\Omega')dEd\Omega.$

Therefore, $T(x+dx,E,\Omega,E_0,\Omega_0)dEd\Omega$ = sum of three terms. Expanding this in a Taylor's series and cancelling appropriate differentials leaves

$$\frac{d}{dx} T(x,E,\Omega,E_0,\Omega_0) = -\Sigma(x,E_0)/|\mu_0| T(x,E,\Omega,E_0,\Omega_0).$$

$$+ \Sigma(x,E_0)c(x,E_0)/|\mu_0| \int_0^\infty dE' \int_0^{4\pi} d\Omega' f(x,E',\Omega',E_0,\Omega_0)$$

$$\cdot T(x,E,\Omega,E',\Omega') + \int_0^\infty dE'' \int_0^{4\pi} d\Omega'' \int_0^\infty dE' \int_0^{4\pi} d\Omega'$$

$$\cdot R(x,E'',\Omega'',E_0,\Omega_0)\Sigma(x,E'')c(x,E'')/\mu'' f(x,E',\Omega',E'',\Omega'')$$

$$\cdot T(x,E,\Omega,E',\Omega'). \tag{6.3-1}$$

Here the initial condition is naturally

$$T(0,E,\Omega,E_0,\Omega_0) = \delta(\Omega - \Omega_0)\,\delta(E - E_0) \tag{6.3-2}$$

When the scattering function has the reduced form of Eq. (3.3-3) so that the ϕ-dependence can be removed by letting.

$$T(x,E,\mu,E_0,\mu_0)dEd\mu = \frac{1}{2\pi} \int_0^{2\pi} d\phi_0 \int_0^{2\pi} d\phi T(x,E,\Omega,E_0,\Omega_0)dEd\mu \tag{6.3-3}$$

along with Eq. (3.3-4), then the resulting equation is

$$\frac{d}{dx} T(x,E,\mu,E_0,\mu_0) = -\Sigma(x,E_0)/|\mu_0| T(x,E,\mu,E_0,\mu_0)$$

$$+ \Sigma(x,E_0)c(x,E_0)/|\mu_0| \int_E^{E_0} dE' \int_{-1}^0 d\mu' f(x,E',E_0,\Omega'\cdot\Omega_0)$$

$$\cdot T(x,E,\mu,E',\mu') + \int_E^{E_0} dE'' \Sigma(x,E'')c(x,E'')$$

$$\cdot \int_0^1 d\mu'' R(x,E''',\mu'',E_0,\mu_0)/\mu'' \int_E^{E''} dE' \int_{-1}^0 d\mu'$$

$$\cdot T(x,E,\mu,E',\mu') f(x,E',E'',\Omega'\cdot\Omega''). \tag{6.3-4}$$

Here the restriction to down scattering in energy has also been applied by the energy integration limits shown.

The resulting transmission because of a distributed surface source then is

$$T(x,E,\mu) = \int_E^{E_0} dE_0 \int_{-1}^0 d\mu_0 \, S(x,E_0,\mu_0) T(x,E,\mu,E_0,\mu_0) \qquad (6.3\text{-}5)$$

while the total transmission is

$$T(x,E_0,\mu_0) = \int_0^{E_0} dE \int_0^1 d\mu \, T(x,E,\mu,E_0,\mu_0). \qquad (6.3\text{-}6)$$

It is noted that an energy transmission can be defined by

$$T_E(x,E,\mu,E_0,\mu_0) = E \, T(x,E,\mu,E_0,\mu_0) \qquad (6.3\text{-}7)$$

which leads naturally to the concept of dose calculations that are a common quantity in the field of radiation shielding.

6.4. FACTORING OF THE TRANSMISSION EQUATION

Because of the delta-function initial condition of the transmission function and the peak in first scattering effects, it is convenient to separate the transmission function in the following manner:

$$T(x,E,\mu,E_0,\mu_0) dE d\mu = T_0(x,E,\mu,E_0,\mu_0) \, \delta(E-E_0) dE \, \delta(\mu-|\mu_0|) d\mu$$
$$(6.4\text{-}1)$$
$$+ T_1(x,E,\mu,E_0,\mu_0) dE d\mu + T_M(x,E,\mu,E_0,\mu_0) dE d\mu,$$

where T_0 is the unscattered component, T_1 is the single scattered component, and T_M is the multiscattered component. Thus Eq. (6.3-4) produces

$$\frac{d}{dx} T_0(x,E,\mu,E_0,\mu_0) \, \delta(E-E_0)\delta(\mu-|\mu_0|) = -\Sigma(x,E_0)T_0(x,E,\mu,E_0,\mu_0)$$
$$(6.4\text{-}2)$$
$$\delta(E-E_0) \, \delta(\mu-|\mu_0|)/|\mu_0|,$$

$$\frac{d}{dx} T_1(x,E,\mu,E_0,\mu_0) = -\Sigma(x,E_0) \, T_1(x,E,\mu,E_0,\mu_0)/|\mu_0|$$
$$(6.4\text{-}3)$$
$$+ \Sigma(x,E_0)c(x,E_0)/|\mu_0| \, f(x,E,E_0,\Omega \cdot \Omega_0) T_0(x,E,\mu,E,\mu),$$

$$\frac{d}{dx} T_M(x,E,\mu,E_0,\mu_0) = -\Sigma(x,E_0) \, T_M(x,E,\mu,E_0,\mu_0)/|\mu_0|$$
$$+ \Sigma(x,E_0)c(x,E_0)/|\mu_0| \int_E^{E_0} dE' \int_{-1}^0 d\mu' \, f(x,E',E_0,\Omega' \cdot \Omega_0)$$

$$\cdot \, [T_1 \, (x,E,\mu,E',\mu') + T_M(x,E,\mu,E',\mu')]$$

$$+ \int_E^{E_0} dE'' \, \Sigma(x,E'')c(x,E'') \int_0^1 d\mu'' [R_1 \, (x,E'',\mu'',E_0,\mu_0)$$

$$+ R_M(x,E'',\mu'',E_0,\mu_0)]/\mu'' \, \{T_0 \, (x,E,\mu,E,\mu) \, f(x,E,E'',\Omega \, \cdot \, \Omega'')$$

$$+ \int_E^{E''} dE' \int_{-1}^0 d\mu' f(x,E',E'',\Omega' \, \cdot \, \Omega'') \, [T_1 \, (x,E,\mu,E',\mu')$$

$$+ T_M \, (x,E,\mu,E',\mu')] \}, \tag{6.4-4}$$

where Eq. (5.3-2) has been utilized for the reflection function.
The initial conditions become

$$T_0 \, (0,E,\mu,E_0,\mu_0) \, \delta(E-E_0) \, \delta(\mu-|\mu_0|) = \delta(E-E_0) \, \delta(\mu-|\mu_0|), \tag{6.4-5}$$

$$T_1 \, (0,E,\mu,E_0,\mu_0) = 0, \tag{6.4-6}$$

and

$$T_M \, (0,E,\mu,E_0,\mu_0) = 0. \tag{6.4-7}$$

The solution to the unscattered component is

$$T_0 \, (x,E,\mu,E,\mu) = \exp \, [- \int_0^x dx' \, \Sigma(x',E)/\mu] \tag{6.4-8}$$

which reduces to

$$T_0 \, (x,E,\mu,E,\mu) = T_0 \, (a,E,\mu,E,\mu) \, \exp \, [\Sigma(E)/\mu \, \{a-x\}] \tag{6.4-9}$$

for a homogeneous layer between $a \leqslant x \leqslant b$. The corresponding results for the single scattered transmission contributions are

$$T_1 \, (x,E,\mu,E_0,\mu_0) = \exp \, [- \int_0^x dx' \Sigma(x,E_0)/|\mu_0|] \int_0^x dx'' \Sigma(x'',E_0)$$

$$\cdot \, c(x'',E_0)/|\mu_0| \, f(x'',E,E_0,\Omega \, \cdot \, \Omega_0) \, \{\exp \, [- \int_0^{x''} dx'$$

$$\cdot \, \Sigma(x',E)/\mu] - \exp\left[-\int_0^{x''} dx' \, \Sigma(x',E_0)/|\mu_0|\right]\}$$ (6.4-10)

and

$$T_1(x,E,\mu,E_0,\mu_0) = T_1(a,E,\mu,E_0,\mu_0) \exp\left[\Sigma(E_0)/|\mu_0| \, \{a-x\}\right]$$ (6.4-11)

$$+ \, T_0(a,E,\mu,E,\mu) \, \Sigma(E_0)c(E_0)/|\mu_0| \, f(E,E_0,\Omega\cdot\Omega_0)[\Sigma(E_0)/|\mu_0|$$

$$- \, \Sigma(E)/\mu]^{-1} \, \{\exp\left[\Sigma(E)/\mu \, \{a-x\}\right] - \exp\left[\Sigma(E_0)/|\mu_0| \, \{a-x\}\right]\}.$$

6.5. APPROXIMATION THEORY

In the case of deep penetration whereby the thickness of the slab becomes large, the reflection function appearing in Eq. (6.3-1) closely approximates that of a semiinfinite medium. If this substitution is made in Eq. (6.3-4), then

$$\frac{d}{dx} T(x,E,\mu,E_0,\mu_0) = -\Sigma(x,E_0)/|\mu_0| T(x,E,\mu,E_0,\mu_0) + \Sigma(x,E_0)c(x,E_0)/|\mu_0|$$

$$\cdot \int_E^{E_0} dE' \int_{-1}^0 d\mu' \, f(x,E',E_0,\Omega\cdot\Omega_0) \, T(x,E,\mu,E',\mu')$$

$$+ \int_E^{E_0} dE'' \, \Sigma(x,E'')c(x,E'') \int_{-1}^0 d\mu'' R(\infty,E'',\mu'',E_0,\mu_0)/\mu''$$

$$\cdot \int_E^{E''} dE' \int_{-1}^0 d\mu' \, T(x,E,\mu,E',\mu')$$

$$\cdot f(x,E',E'',\Omega'\cdot\Omega'') \qquad (x \gg 0).$$ (6.5-1)

This equation is now mathematically linear in form although still of integro-differential in type. Further use will be made of this approximation in later chapters.

6.6. DISCRETE FORMULATION

In comparison with the reflection function, the discrete transmission is indicated by

$$T_{mn}^{ij}(x) = T(x, E_m, \mu_i, E_n, \mu_j) \tag{6.6-1}$$

so that the results are written for the ϕ-independent case for layered homogeneous slabs by

$$T_{mm0}^{ii}(x) = T_{mm0}^{ii}(a) \exp\left[\Sigma_m/\mu_i \{a-x\}\right], \tag{6.6-2}$$

$$T_{mn1}^{ij}(x) = T_{mn1}^{ij}(a) \exp\left[\Sigma_n/\mu_j \{a-x\}\right] + T_{mm0}^{ii}(a) \, \Sigma_n c_n/\mu_j \, f_{mn}^{ij} \left[\Sigma_n/\mu_j\right.$$

$$\left. - \Sigma_m/\mu_i\right]^{-1} \left[\exp\left[\Sigma_m/\mu_i \{a-x\}\right] - \exp\left[\Sigma_n/\mu_j \{a-x\}\right]\right], \tag{6.6-3}$$

and

$$\frac{d}{dx} T_{mnM}^{ij}(x) = -\Sigma_n T_{mnM}^{ij}(x)/\mu_j + \Sigma_n c_n/\mu_j \sum_{p=1}^{N} \Delta E_p \sum_{s=1}^{L} w_s \, f_{pn}^{sj} \left[T_{mp1}^{is}(x)\right.$$

$$\left. + T_{mpM}^{is}(x)\right] + \sum_{q=1}^{N} \Delta E_q \, \Sigma_q c_q \sum_{u=1}^{L} w_u \, \left[R_{qnl}^{uj}(x) + R_{qnM}^{uj}(x)\right]/\mu_u \{T_{mm0}^{ii}(x)$$

$$\cdot \hat{f}_{mq}^{iu} + \sum_{p=1}^{q} \Delta E_p \sum_{s=1}^{L} w_s \hat{f}_{pq}^{su} \left[T_{mp1}^{is}(x) + T_{mpM}^{is}(x)\right]\}, \tag{6.6-4}$$

with the initial condition

$$T_{mnM}^{ij}(0) = 0. \tag{6.6-5}$$

Here $i,j = 1,2,\ldots,L; m,n = 1,2,\ldots,N$.

Naturally, the other functional equations for the transmission function can be discretized also; however, it is straightforward to do so and will not be listed here. Equations (6.6-2,3,4) are the normally utilized versions.

The modified form of the transmission function is produced by letting

$$\Im_{mn}^{ij}(x) = T_{mn}^{ij}(x) \, \Delta E_m \tag{6.6-6}$$

and using Eqs. (6.3-13,14) produces

$$\frac{d}{dx} \Im_{mnM}^{ij}(x) = -\Sigma_n \Im_{mnM}^{ij}(x)/\mu_j + \Sigma_n c_n/\mu_j \sum_{p=1}^{N} \sum_{s=1}^{L} w_s \, f_{pn}^{sj} \, \Im_{mpS}^{is}(x)$$

$$+ \sum_{q=1}^{N} \Sigma_q c_q \sum_{u=1}^{L} w_u \mathcal{R}_{qn}^{uj}(x)/\mu_u \; \{\mathcal{I}_{mn0}^{ii}(x) \; \hat{l}_{mq}^{iu}$$

$$+ \sum_{p=1}^{q} \sum_{s=1}^{L} w_s \; \hat{l}_{pq}^{su} \; \mathcal{I}_{mpS}^{is}(x)\},$$ (6.6-7)

where

$$\mathcal{I}_{mnS}^{ij}(x) = \mathcal{I}_{mn1}^{ij}(x) + \mathcal{I}_{mnM}^{ij}(x).$$ (6.6-8)

The analogous equations for the unscattered and first scattered transmitted currents can be written down in a like manner.

6.7. A SPECIAL CASE

The important special use of monoenergetic transport can be obtained by utilizing

$$T(x,\mu,\mu_0)d\mu = \int_0^{\infty} dE_0 T(x,E,\mu,E_0,\mu\ )d\mu\ \delta(E-E_0)$$ (6.7-1)

producing the result

$$\frac{d}{dx} T(x,\mu,\mu_0) = - \Sigma(x)/|\mu_0| T(x,\mu,\mu_0)$$

$$+ \Sigma(x)c(x)/|\mu_0| \int_{-1}^{0} d\mu' \; f(x,\mu',\mu_0) \; T(x,\mu,\mu')$$

$$+ \Sigma(x)c(x) \int_0^1 d\mu'' R(x,\mu'',\mu\ )/\mu'' \int_{-1}^{0} d\mu' f(x,\mu',\mu'')$$

$$\cdot T(x,\mu,\mu').$$ (6.7-2)

By utilizing the mean free path, Eq. (3.4-5), and the transformation

$$\hat{T}(x,\mu,\mu_0) = \mu_0 \; T(x,\mu,\mu_0),$$ (6.7-3)

where all angles are defined with respect to their respective normals, a symmetric form for the scattered transmission is obtained when isotropic scattering, Eq. (3.4-6) is employed as

$$\frac{d}{dx} \hat{T}_s (\tilde{x},\mu,\mu_0) = -\hat{T}_s (\tilde{x},\mu,\mu_0)/\mu_0 + c/2 \left[1 + \int_0^1 d\mu'' \hat{R}(\tilde{x},\mu'',\mu_0)/\mu'' \right]$$

$$\left[\hat{T}_0 (\tilde{x},\mu,\mu)/\mu + \int_0^1 d\mu' \, \hat{T}_s (\tilde{x},\mu,\mu')/\mu' \right], \qquad (6.7\text{-}4)$$

where

$$\hat{T}(\tilde{x},\mu,\mu_0) = \hat{T}_0(\tilde{x},\mu_0,\mu_0) + \hat{T}_s (\tilde{x},\mu,\mu_0) \qquad (6.7\text{-}5)$$

and

$$\hat{T}_0(\tilde{x},\mu_0,\mu_0) = \mu_0 \exp \left[- \tilde{x}/\mu_0 \right]. \qquad (6.7\text{-}6)$$

Again the analogous results for the reflection function have been employed.

The discrete forms of this result corresponding to the homogeneous layered slab model are

$$\frac{d}{d\tilde{x}} \hat{T}_s^{ij} (\tilde{x}) = -\hat{T}_s^{ij}(\tilde{x})/\mu_j + c/2 \, \Gamma_j [\hat{T}_0^{ii} (\tilde{x})/\mu_i + \sum_{l=1}^{L} w_l \hat{T}_s^{il} (\tilde{x})/\mu_l], \qquad (6.7\text{-}7)$$

where

$$\hat{T}_s^{ij} (\tilde{x}) = \mu_j [T_1^{ij} (\tilde{x}) + T_M^{ij}(\tilde{x})], \qquad (6.7\text{-}8)$$

$$\hat{T}_0^{ii} (\tilde{x}) = \mu_i \exp [-\tilde{x}/\mu_i], \qquad (6.7\text{-}9)$$

and

$$\Gamma_j = \Gamma(\mu_j) = 1 + \sum_{l=1}^{L} w_l \, \hat{R}^{lj} (\tilde{x})/\mu_l. \qquad (6.7\text{-}10)$$

Further uses of this monoenergetic case will be shown in a later chapter.

6.8. COMPUTATIONAL CONSIDERATIONS

Certain computational considerations are discussed in the appendix since they are common to all cases. However, it is to be noted that for the mono-energetic special case the system is small enough that the separation of the first and multiscattered contributions to the transmitted current is no longer necessary and just the unscattered and scattered contributions are considered separately. On the other hand, most energy dependent problems where the scattering is nonisotropic in nature, will require the full decomposition of the transmission function.

6.9. EXERCISES

1. Derive Eq. (6.3-1) by starting with the basic definitions in the form of differences, i.e., $T(x + \Delta x, E, \Omega, E_0, \Omega_0) \Delta E \Delta \Omega$, and take the limit as $\Delta x \to 0$.

2. Fill in the steps in going from Eq. (6.3-2) to Eq. (6.3-4).

3. Determine the relationship between Chandrasekhar's (1950) X- and Y-functions and the R- and T-functions utilized here.

4. Derive the transmission function by adding a thickness dx on the left face, i.e., the particle emerging side, of a slab by the particle counting procedure.

5. Show that Eq. (6.4-11) reduces to zero for the special case of $\mu = |\mu_0|$.

CHAPTER 7

ESCAPE FUNCTION CONCEPTS

7.1. INTRODUCTION

Another class of problems that arise in transport theory are concerned with the calculations involving internal sources. In order to handle these type problems, the concept of the escape probability for a medium is introduced.

7.2. DEFINITIONS

The following escape function is defined:

$E_R(x,E,\Omega)dEd\Omega$ = the particle current* escaping from the right face of a slab of thickness x in the energy range dE about E in the direction $d\Omega$ about Ω because of the presence of an internal volumetric source $S(z,E_s,\Omega_s)$ where $0 \leqslant z \leqslant x$.

However, this escape function is intimately dependent upon the source distribution; therefore, a more basic quantity is the escape Green's function. This is:

$G_R(x,E,\Omega,z,E_0,\Omega_0)dEd\Omega$ = the particle current* escaping from the right face of a slab of thickness x in the energy range dE about E in the direction $d\Omega$ about Ω because of the presence of a unit internal plane source of energy E_0 and direction Ω_0 at z where $0 \leqslant z \leqslant x$.

From this Green's function definition then

$$E_R(x,E,\Omega) = \int_0^x dz \int_0^\infty dE_0 \int_0^{4\pi} d\Omega_0\, S(z,E_0,\Omega_0)$$

$$\cdot\; G_R(x,E,\Omega,z,E_0,\Omega_0)$$

(7.2-1)

---

*Current is explained in Section 3.2.

The choice as to whether to use the G_R or E_R function generally depends upon the nature of the internal source. If the source is a known and fixed quantity, then the E_R function is most useful; on the other hand, if the source can vary from case to case, then the G_R function is the best to use.

Naturally, similar quantities referring to escape from the left face of the slab can be defined by using a L subscript.

7.3. DERIVATION OF THE ESCAPE FUNCTIONS

The derivation of the escape functions proceeds along the same lines as the reflection and transmission functions. In reference to Fig. 7 the terms are:

1. Source in dx emits to the right.

2. Source in dx emits to the left, reflection; no interaction on leaving.

3. Escape from x, no interaction on leaving.

4. Escape from x, interaction leaving with scatter outward.

5. Escape from x, interaction leaving with scatter inward, reflection; no interaction leaving.

The mathematical terms are:

1. $S(x,E,\Omega)dEd\Omega dx$,

2. $\int_0^\infty \int_{2\pi}^{4\pi} S(x,E',\Omega')dE'd\Omega'dx\, R(x,E,\Omega,E',\Omega')dEd\Omega[1-\Sigma(x,E)dx/\mu]$,

3. $E_R(x,E,\Omega)dEd\Omega[1-\Sigma(x,E)dx/\mu]$,

4. $\int_0^\infty \int_0^{2\pi} E_R(x,E'',\Omega'')dE''d\Omega''\,\Sigma(x,E'')dx/\mu''\,c(x,E'')$

$\cdot f(x,E,\Omega,E'',\Omega'')dEd\Omega$,

5. $\int_0^\infty \int_0^{2\pi} \int_0^\infty \int_{2\pi}^{4\pi} E_R(x,E'',\Omega'')dE''d\Omega''\,\Sigma(x,E'')dx/\mu''\,c(x,E'')$

$\cdot f(x,E',\Omega',E'',\Omega'')dE'd\Omega'\,R(x,E,\Omega,E',\Omega')dEd\Omega[1-\Sigma(x,E)dx/\mu]$.

The usual Taylor's series expansion of $E_R(x+dx)$ set equal to the sum of the five terms produces the differential equation

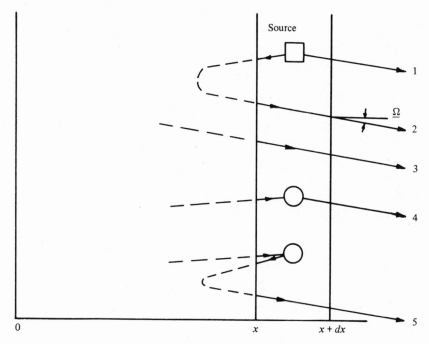

Fig. 7. The escape function interactions.

$$\frac{d}{dx} E_R(x,E,\Omega) = - \Sigma(x,E)/\mu E_R(x,E,\Omega) + S(x,E,\Omega)$$

$$+ \int_E^{E_0} dE' \int_{2\pi}^{4\pi} d\Omega' S(x,E',\Omega') R(x,E,\Omega,E',\Omega') + \int_E^{E_0} dE'' \int_0^{2\pi} d\Omega''$$

$$\cdot \Sigma(x,E'')c(x,E'')/\mu'' E_R(x,E'',\Omega'') \{f(x,E,\Omega,E'',\Omega'')$$

$$+ \int_E^{E''} dE' \int_{2\pi}^{4\pi} d\Omega' f(x,E',\Omega',E'',\Omega'') R(x,E,\Omega,E',\Omega')\} . \qquad (7.3\text{-}1)$$

The initial condition with no external surface source is naturally

$$E_R(0,E,\Omega) = 0. \qquad (7.3\text{-}2)$$

It is to be noted that this escape function can represent the transmission of a surface source through the slab if the internal source is set equal to zero and if the boundary condition

$$E_R(0,E,\Omega) = \delta(E - E_0)\,\delta(\Omega - \Omega_0) \qquad (7.3\text{-}3)$$

is imposed. In some applications it is convenient to utilize this escape function as both the result of an internal source and a distributed input to the slab.

If now the source is taken as delta-functions

$$S(x,E,\Omega) = \delta(x - z)\,\delta(E - E_0)\,\delta(\Omega - \Omega_0), \qquad (7.3\text{-}4)$$

then the E_R function becomes the G_R function. The differential equation then is

$$\frac{d}{dx}\,G_R(x,E,\Omega,z,E_0,\Omega_0) = -\,\Sigma(x,E)/\mu\; G_R(x,E,\Omega,z,E_0,\Omega_0) + \delta(x - z)$$

$$\cdot\;\{\delta(E - E_0)\,\delta(\Omega - \Omega_0) + R(x,E,\Omega,E_0,\Omega_0)\} + \int_E^{E_0} dE''\int_0^{2\pi} d\Omega''$$

$$\cdot\;\Sigma(x,E'')\,c(x,E'')/\mu''\; G_R(x,E'',\Omega'',z,E_0,\Omega_0)\;\{f(x,E,\Omega,E'',\Omega'')$$

$$+ \int_E^{E''} dE'\int_{2\pi}^{4\pi} d\Omega'\, f(x,E',\Omega',E'',\Omega'')\,R(x,E,\Omega,E',\Omega')\} \qquad (7.3\text{-}5)$$

with

$$G_R(0,E,\Omega,z,E_0,\Omega_0) = 0. \qquad (7.3\text{-}6)$$

Naturally, this can be rewritten as

$$\frac{d}{dx}\,G_R(x,E,\Omega,z,E_0,\Omega_0) = -\,\Sigma(x,E)/\mu\; G_R(x,E,\Omega,z,E_0,\Omega_0)$$

$$+ \int_E^{E_0} dE''\int_0^{2\pi} d\Omega''\,\Sigma(x,E'')\,c(x,E'')/\mu''\; G_R(x,E'',\Omega'',z,E_0,\Omega_0)$$

$$\cdot\;\{f(x,E,\Omega,E'',\Omega'') + \int_E^{E''} dE'\int_{2\pi}^{4\pi} d\Omega'\, f(x,E',\Omega',E'',\Omega'')$$

$$\cdot\;R(x,E,\Omega,E',\Omega')\} \qquad x > z, \qquad (7.3\text{-}7)$$

with the boundary condition

$$G_R(z,E,\Omega,z,E_0,\Omega_0) = \begin{cases} \delta(E - E_0)\delta(\Omega - \Omega_0) & 0 \leqslant \Omega_0 < 2\pi \\ \\ R(z,E,\Omega,E_0,\Omega_0) & 2\pi < \Omega_0 \leqslant 4\pi \end{cases} \qquad (7.3\text{-}8)$$

and where

$$G_R(x,E,\Omega,z,E_0,\Omega_0) = 0 \qquad x < z. \qquad (7.3\text{-}9)$$

The reduction of these equations to the case of ϕ-independence follows naturally by the transformations

$$E_R(x,E,\mu)\,d\mu = \int_0^{2\pi} d\phi\, E_R(x,E,\Omega)d\mu, \qquad (7.3\text{-}10)$$

$$G_R(x,E,\mu,z,E_0,\mu_0)\,d\mu = \frac{1}{2\pi}\int_0^{2\pi}d\phi_0 \int_0^{2\pi}d\phi\, G_R(x,E,\Omega,z,E_0,\Omega_0)\,d\mu, \qquad (7.3\text{-}11)$$

$$S(x,E,\mu)\,d\mu = \frac{1}{2\pi}\int_0^{2\pi}d\phi\, S(x,E,\Omega)\,d\mu. \qquad (7.3\text{-}12)$$

The resulting differential equations are

$$\frac{d}{dx}E_R(x,E,\mu) = -\Sigma(x,E)/\mu\,E_R(x,E,\mu) + S(x,E,\mu)$$

$$+ \int_E^{E_0}dE' \int_{-1}^{0} d\mu'\, S(x,E',\mu')\, R(x,E,\mu,E',\mu') + \int_E^{E_0}dE'' \int_0^1 d\mu''$$

$$\cdot \Sigma(x,E'')\,c(x,E'')/\mu''\, E_R(x,E'',\mu'')\, \{f(x,E,E'',\Omega \cdot \Omega'')$$

$$+ \int_E^{E_0}dE' \int_{-1}^{0} d\mu'\, f(x,E',E'',\Omega' \cdot \Omega'')\, R(x,E,\mu,E',\mu')\}, \qquad (7.3\text{-}13)$$

and

$$\frac{d}{dx}G_R(x,E,\mu,z,E_0,\mu_0) = -\Sigma(x,E)/\mu\, G_R(x,E,\mu,z,E_0,\mu_0) + \int_E^{E_0}dE'' \int_0^1 d\mu''$$

$$\cdot c(x,E'')\,\Sigma(x,E'')/\mu''\, G_R(x,E'',\mu'',z,E_0,\mu_0)\,\{f(x,E,E'',\Omega \cdot \Omega'')$$

$$+ \int_E^{E''}dE' \int_{-1}^{0} d\mu'\, f(x,E',E'',\Omega' \cdot \Omega'')\, R(x,E,\mu,E',\mu')\} \qquad x \geqslant z \qquad (7.3\text{-}14)$$

with the boundary conditions

$$E_R(0,E,\mu) = 0,$$

$$G_R(z,E,\mu,z,E_0,\mu_0) = \begin{cases} \delta(E-E_0)\,\delta(\mu-\mu_0) & \mu_0 > 0 \\ R(z,E,\mu,E_0,\mu_0) & \mu_0 < 0, \end{cases} \qquad (7.3\text{-}15)$$

and implying that

$$G_R(x,E,\mu,z,E_0,\mu_0) = 0 \qquad x < z. \qquad (7.3\text{-}16)$$

Equation (7.3-14) is a modified transmission-type function with specific boundary conditions.

7.4. FACTORING OF THE ESCAPE FORMULATION

In theory the same concepts can be employed of factoring the escape function, E_R, into unscattered, single-scattered and multiscattered components; however, because of their inherent dependence upon the source function, it is not of any computational advantage to perform this factoring. On the other hand, it is convenient to factor the Green's function, G_R. Thus let

$$G_R(x,E,\mu,E_0,\mu_0) = G_{R0}(x,E,\mu,z,E_0,\mu_0)\,\delta(E-E_0)\,\delta(\mu-|\mu_0|)$$

$$+ G_{R1}(x,E,\mu,z,E_0,\mu_0) + G_{RM}(x,E,\mu,z,E_0,\mu_0) \qquad (7.4\text{-}1)$$

producing the differential equations

$$\frac{d}{dx}\,G_{R0}(x,E,\mu,z,E_0,\mu_0)\,\delta(E-E_0)\,\delta(\mu-|\mu_0|) = -\,\Sigma(x,E)/\mu$$

$$\cdot\,G_{R0}(x,E,\mu,z,E_0,\mu_0)\,\delta(E-E_0)\,\delta(\mu-|\mu_0|), \qquad (7.4\text{-}2)$$

$$\frac{d}{dx}\,G_{R1}(x,E,\mu,z,E_0,\mu_0) = -\,\Sigma(x,E)/\mu\,G_{R1}(x,E,\mu,z,E_0,\mu_0)$$

$$+ \Sigma(x,E_0)\,c(x,E_0)/|\mu_0|G_{R0}(x,E_0,\mu_0,z,E_0,\mu_0)\,f(x,E,E_0,\Omega\cdot\Omega_0), \qquad (7.4\text{-}3)$$

and

$$\frac{d}{dx}\,G_{RM}(x,E,\mu,z,E_0,\mu_0) = -\,\Sigma(x,E)/\mu\,G_{RM}(x,E,\mu,z,E_0,\mu_0)$$

$$+ \int_E^{E_0} dE''\int_0^1 d\mu''\,\Sigma(x,E'')\,c(x,E'')/\mu''\,\{G_{R1}(x,E'',\mu'',z,E_0,\mu_0)$$

$$\cdot f(x,E,E'',\Omega \cdot \Omega'') + G_{RM}(x,E'',\mu'',z,E_0,\mu_0) \ \{f(x,E,E'',\Omega \cdot \Omega'')$$

$$+ \int_E^{E''} dE' \int_{-1}^0 d\mu' \, f(x,E',E'',\Omega' \cdot \Omega'') \ [R_1(x,E,\mu,E',\mu')$$

$$+ R_M(x,E,\mu,E',\mu')] \ \} \ \}, \tag{7.4-4}$$

with the boundary conditions

$$G_{R0}(z,E,\mu,z,E_0,\mu_0)\delta(E - E_0) \, \delta(\mu - |\mu_0|) = \begin{cases} \delta(E - E_0) \, \delta(\mu - \mu_0) & \mu_0 > 0 \\ 0 & \mu_0 < 0, \end{cases}$$

$$\tag{7.4-5}$$

$$G_{R1}(z,E,\mu,z,E_0,\mu_0) = \begin{cases} 0 & \mu_0 > 0 \\ R_1(z,E,\mu,E_0,\mu_0) & \mu_0 < 0, \end{cases} \tag{7.4-6}$$

and

$$G_{RM}(z,E,\mu,z,E_0,\mu_0) = \begin{cases} 0 & \mu_0 > 0 \\ R_M(z,E,\mu,E_0,\mu_0) & \mu_0 < 0. \end{cases} \tag{7.4-7}$$

This different boundary condition for inwardly or outwardly directed sources leads to the further separation of the Green's function into

$$G_R^+(x,E,\mu,z,E_0,\mu_0) = G_R(x,E,\mu,z,E_0,\mu_0) \qquad \mu_0 > 0 \tag{7.4-8}$$

and

$$G_R^-(x,E,\mu,z,E_0,\mu_0) = G_R(x,E,\mu,z,E_0,\mu_0) \qquad \mu_0 < 0. \tag{7.4-9}$$

These results can then be written as

$$G_{R0}^+(x,E_0,\mu_0,z,E_0,\mu_0) = \exp\left[-\int_z^x dx' \, \Sigma(x',E_0)/|\mu_0|\right]$$

$$= T_0(x-z,E_0,\mu_0,E_0,\mu_0), \tag{7.4-10}$$

$$G_{R0}^-(x,E_0,\mu_0,z,E_0,\mu_0) = 0, \tag{7.4-11}$$

$$G_{R1}^+(x,E,\mu,z,E_0,\mu_0) = \exp\left[-\int_z^x dx' \, \Sigma(x',E)/\mu\right] \int_z^x dx'' \, \Sigma(x'',E_0)$$

$$\cdot \, c(x'',E_0)/|\mu_0| f(x'',E,E_0,\Omega \cdot \Omega_0) \exp\left[\int_z^{x''} dx' \, \Sigma(x',E)/\mu\right]$$

$$\cdot \exp \left[- \int_z^{x''} dx' \ \Sigma(x',E_0)/|\mu_0| \right] = T_1(x-z,E,\mu,E_0,\mu_0), \qquad (7.4\text{-}12)$$

and

$$G_{R1}^-(x,E,\mu,z,E_0,\mu_0) = R_1(z,E,\mu,E_0,\mu_0) \exp \left[- \int_0^x dx' \ \Sigma(x',E)/\mu \right]. \quad (7.4\text{-}13)$$

The relationships to the standard transmissions functions are also noted. In the restricted case of a layered homogeneous slab the result is

$$G_{R0}^+(x,E_0,\mu_0,z,E_0,\mu_0) = G_{R0}^+(a,E_0,\mu_0,z,E_0,\mu_0) \exp \left[\Sigma(E_0)/|\mu_0| \ \{a-x\} \right], (7.4\text{-}14)$$

$$G_{R1}^+(x,E,\mu,z,E_0,\mu_0) = G_{R1}^+(a,E,\mu,z,E_0,\mu_0) \exp \left[\Sigma(E)/\mu \ \{a-x\} \right]$$

$$+ \Sigma(E_0)c(E_0)/|\mu_0| G_{R0}^+(a,E_0,\mu_0,z,E_0,\mu_0)f(E,E_0,\Omega \cdot \Omega_0) \ \{\Sigma(E)/\mu$$

$$- \Sigma(E_0)/|\mu_0|\}^{-1} \{\exp \left[\Sigma(E_0)/|\mu_0| \ \{a-x\} \right] - \exp\left[\Sigma(E)/\mu\{a-x\}\right]\} \}, \quad (7.4\text{-}15)$$

$$G_{R1}^-(x,E,\mu,z,E_0,\mu_0) = G_{R1}^-(a,E,\mu,z,E_0,\mu_0) \exp \left[\Sigma(E)/\mu\{a-x\} \right], \qquad (7.4\text{-}16)$$

and

$$G_{R1}^-(z,E,\mu,z,E_0,\mu_0) \ = \ R_1(z,E,\mu,E_0,\mu_0) \qquad\qquad (7.4\text{-}17)$$

In all these results $x \geqslant a \geqslant z$ and all answers are identically zero if $x < z$.

7.5. THE DISCRETE FORMULATION

The transformation of these equations into discrete forms follows from the previous results.

$$G_{Rmn}^{\pm ij}(x,z) = G_R^\pm(x,E_m,\mu_i,z,E_n,\mu_j) \qquad\qquad (7.5\text{-}1)$$

and

$$E_{Rm}^i(x) = E_R(x,E_m,\mu_i). \qquad\qquad (7.5\text{-}2)$$

Then Equation (7.3-13) has the discrete form

$$\frac{d}{dx}E_{Rm}^i(x) = - \Sigma_m(x)/\mu_i E_{Rm}^i(x) + S_m^i(x) + \sum_{p=1}^{N} \Delta E_p \sum_{s=1}^{L} w_s$$

$$\cdot S_p^s(x)[R_{mpI}^{is}(x) + R_{mpn}^{is}(x)] + \sum_{q=1}^{N} \Delta E_q \sum_{u=1}^{L} w_u \Sigma_q(x) c_q(x)/\mu_u$$

$$\cdot E_{Rq}^u(x) \{ f_{mq}^{iu}(x) + \sum_{p=1}^{N} \Delta E_p \sum_{s=1}^{L} w_s \hat{f}_{pq}^{su}(x) [R_{mpI}^{is}(x) + R_{mpM}^{is}(x)] \}. \tag{7.5-3}$$

The corresponding results for the homogeneous layered slab is

$$G_{Rmm0}^{+jj}(x,z) = G_{Rmm0}^{+jj}(a,z) \exp [\Sigma_m/\mu_j \{a-x\}], \tag{7.5-4}$$

$$G_{Rmn1}^{+ij}(x,z) = G_{Rmn1}^{+ij}(a,z) \exp [\Sigma_m/\mu_i \{a-x\}] + \Sigma_n c_n/\mu_j$$

$$\cdot G_{Rmn0}^{+jj}(a,z) \{\Sigma_m/\mu_i - \Sigma_n/\mu_j\}^{-1} f_{mn}^{ij} \{\exp [\Sigma_n/\mu_j \{a-x\}]$$

$$- \exp [\Sigma_m/\mu_i \{a-x\}]\}, \tag{7.5-5}$$

$$G_{Rmn1}^{-ij}(x,z) = G_{Rmn1}^{-ij}(a,z) \exp [\Sigma_m/\mu_i \{a-x\}], \tag{7.5-6}$$

$$G_{Rmn1}^{-ij}(z,z) = R_{mn1}^{ij}(z), \tag{7.5-7}$$

with the differential equation

$$\frac{d}{dx} G_{RmnM}^{\pm ij}(x,z) = - \Sigma_m/\mu_i G_{RmnM}^{\pm ij}(x,z) + \sum_{q=1}^{N} \Delta E_q \sum_{u=1}^{L} w_u$$

$$\cdot \Sigma_q c_q/\mu_u \{G_{Rqn1}^{\pm uj}(x,z) f_{mq}^{iu} + G_{RqnM}^{\pm uj}(x,z) \{f_{mq}^{iu} + \sum_{p=1}^{q} \Delta E_p \sum_{s=1}^{L} w_s \hat{f}_{pq}^{su}$$

$$\cdot [R_{mp1}^{is}(x) + R_{mpM}^{is}(x)]\} \} \tag{7.5-8}$$

with

$$G_{RmnM}^{+ij}(z,z) = 0 \tag{7.5-9}$$

and

$$G_{RmnM}^{-ij}(z,z) = R_{mnM}^{ij}(z). \tag{7.5-10}$$

In this discrete representation $i, j = 1, 2, \ldots, L; m, n = 1, 2, \ldots, N$. The modified form of the escape functions become

$$\mathcal{E}_{Rm}^i(z) = E_{Rm}^i(x) \Delta E_m$$

and

$$\mathcal{G}_{\notin Rmn}^{\pm ij}(x,z) = G_{Rmn}^{\pm ij}(x,z)\Delta E_m \qquad (7.5\text{-}11)$$

producing

$$\frac{d}{dx}\,\mathcal{E}_{Rm}^i(x) = -\,\Sigma_m(x)/\mu_i\,\mathcal{E}_{Rm}^i(x) + \mathcal{S}_m^i(x)$$

$$+ \sum_{p=1}^{N}\,\sum_{s=1}^{L}\,w_s\,\mathcal{S}_p^s(x)\,\mathcal{R}_{mp}^{is}(x)$$

$$+ \sum_{q=1}^{N}\,\sum_{u=1}^{L}\,w_u\Sigma_q(x)\,c_q(x)/\mu_u\,\mathcal{E}_{Rq}^u(x)\,[\,\ell_{mq}^{iu}(x)$$

$$+ \sum_{p=1}^{N}\,\sum_{s=1}^{L}\,w_s\hat{\ell}_{pq}^{su}(x)\,\mathcal{R}_{mp}^{is}(x)] \qquad (7.5\text{-}12)$$

and

$$\frac{d}{dx}\,\mathcal{G}_{\notin Rmn}^{\pm ij}(x,z) = -\,\Sigma_m(x)/\mu\,\mathcal{G}_{\notin Rmn}^{\pm ij}(x,z)$$

$$+ \sum_{q=1}^{N}\,\sum_{u=1}^{L}\,w_s\,\Sigma_q(x)\,c_q(x)/\mu_u\,\mathcal{G}_{\notin Rqn}^{\pm uj}(x,z)\,[\,\hat{\ell}_{mq}^{iu}(x)$$

$$+ \sum_{p=1}^{N}\,\sum_{s=1}^{L}\,w_s\hat{\ell}_{pq}^{su}(x)\,\mathcal{R}_{mp}^{is}(x)]. \qquad (7.5\text{-}13)$$

Again the usual scattering matrix change is necessary. The discrete modified form in terms of the orders of scattering follows naturally and will not be shown here.

7.6. A SPECIAL CASE

The reduction of the energy dependent theory to that of the monoenergetic case proceeds by using Eq. (3.4-3) and Eq. (6.7-1) along with

$$E_R(x,\mu) = \int_0^\infty dE\,E_R(x,E,\mu)\,d\mu\,\delta(E - E_0), \qquad (7.6\text{-}1)$$

where the monoenergetic energy is E_0 and is dropped from the notation for convenience. The results are

$$\frac{d}{dx}\,E_R(x,\mu) = -\,\Sigma(x)/\mu\,E_R(x,\mu) + S(x,\mu)$$

$$+ \int_{-1}^{0} d\mu' \, S(x, \mu') \, R(x, \mu, \mu')$$

$$+ \Sigma(x) \, c(x) \int_{0}^{1} d\mu'' / \mu'' \, E_R(x, \mu'') \, \{f(x, \mu, \mu')$$

$$+ \int_{-1}^{0} d\mu' \, f(x, \mu', \mu'') \, R(x, \mu, \mu')\}.$$

$$(7.6\text{-}2)$$

The corresponding result for the Green's function is

$$\frac{d}{dx} \, G_R(x, \mu, z, \mu_0) = -\Sigma(x)/\mu \; G_R(x, \mu, z, \mu_0)$$

$$+ \Sigma(x) \, c(x) \int_{0}^{1} d\mu'' G_R(x, \mu'', z, \mu_0) / \mu'' \, [f(x, \mu, \mu'')$$

$$+ \int_{-1}^{0} d\mu' \, f(x, \mu', \mu'') \, R(x, \mu, \mu')].$$

$$(7.6\text{-}3)$$

The discrete forms of these equations follow naturally and will be considered more fully in Chapter 8.

7.7. EXERCISES

1. Derive the equations for the monoenergetic total escape function

$$E_T(x) = \int_{0}^{1} d\mu \, E_R(x, \mu) + \int_{-1}^{0} d\mu \, E_L(x, \mu)$$

for a uniform isotropic source and solve analytically for the unscattered and single scattered components for a homogeneous slab.

2. Determine the relationship between the total escape function for a uniform isotropic source and the "Blackness Coefficient."

$$\beta(x) = 1 - 2[\int_{0}^{1} d\mu \int_{-1}^{0} d\mu_0 \, |\mu_0| \, R(x, \mu, \mu_0)$$

$$+ \int_{-1}^{0} d\mu \int_{-1}^{0} d\mu_0 \, |\mu_0| \, T(x, \mu, \mu_0)]$$

Physically what does the "Blackness Coefficient" represent?

3. Determine the unscattered, single-scattered, and multiscattered forms of Eqs. (7.5-12, 13).

CHAPTER 8

THE MONOENERGETIC CASE

8.1. INTRODUCTION

The monoenergetic case, which is sometimes referred to as the energy-independent case, is an important special result for invariant imbedding transport theory. Many physical problems can, under plausible assumptions, be reduced to a single energy group and still give valuable information. In a physical sense the concept of a "thermal group" of neutrons occurs when thermal equilibrium is assumed between the media and the particles. In practice this equilibrium is only approached because absorption and particle leakage violate its definition; however, in many instances it is possible to model effectively many nuclear reactor subsystems by using a monoenergetic thermal model. Physical properties for this model, like cross sections for instance, should be determined by an energy average over the neutron flux; however, in general this degree of detail is unknown and experimental measurements are made directly using a thermal group of neutrons.

No corresponding case exists for gamma rays since the losses by photoelectric absorption become very large at low photon energies. However, this concept is applicable to electron transport.

8.2. REDUCTION TO THE MONOENERGETIC EQUATIONS

The general energy-dependent invariant imbedding transport theory is reduced to the monoenergetic form in Sections 3.4, 6.7, and 7.6. For convenience these equations are repeated with the nomenclature change of using ξ for the mean free paths rather than the $\tilde{x}$ of Eq. (3.4-5). Thus

$$\frac{d}{d\xi} R(\xi, \mu, \mu_0) = - \left[1/\mu + 1/|\mu_0|\right] R(\xi, \mu, \mu_0)$$

$$+ c(\xi) f(\xi, \mu, \mu_0)/|\mu_0| + c(\xi)/|\mu_0| \int_{-1}^{0} d\mu' f(\xi, \mu', \mu_0) R(\xi, \mu, \mu')$$

$$+ c(\xi) \int_{0}^{1} d\mu'' R(\xi, \mu'', \mu_0) f(\xi, \mu, \mu'')/\mu''$$

$$+ c(\xi) \int_{0}^{1} d\mu'' R(\xi, \mu'', \mu_0)/\mu'' \int_{-1}^{0} d\mu' f(\xi, \mu', \mu'') R(\xi, \mu, \mu'),$$

$$(8.2\text{-}1)$$

$$\frac{d}{d\xi} T(\xi,\mu,\mu_0) = - T(\xi,\mu,\mu_0)/|\mu_0|$$

$$+ c(\xi)/|\mu_0| \int_{-1}^{0} d\mu' f(\xi,\mu',\mu_0) T(\xi,\mu,\mu')$$

$$+ c(\xi) \int_{0}^{1} d\mu'' R(\xi,\mu'',\mu_0)/\mu'' \int_{-1}^{0} d\mu' f(\xi,\mu',\mu'') T(\xi,\mu,\mu'),$$

$$(8.2\text{-}2)$$

$$\frac{d}{d\xi} E_R(\xi,\mu) = - E_R(\xi,\mu)/\mu + S(\xi,\mu) + \int_{-1}^{0} d\mu' S(\xi,\mu') R(\xi,\mu,\mu')$$

$$+ c(\xi) \int_{0}^{1} d\mu'' E_R(\xi,\mu'')/\mu'' [f(\xi,\mu,\mu'')$$

$$+ \int_{-1}^{0} d\mu' f(\xi,\mu',\mu'') R(\xi,\mu,\mu')],$$

$$(8.2\text{-}3)$$

and

$$\frac{d}{d\xi} G_R(\xi,\mu,\eta,\mu_0) = - G_R(\xi,\mu,\eta,\mu_0)/\mu$$

$$+ c(\xi) \int_{0}^{1} d\mu'' G_R(\xi,\mu'',\eta,\mu_0)/\mu'' [f(\xi,\mu,\mu'')$$

$$+ \int_{-1}^{0} d\mu' f(\xi,\mu',\mu'') R(\xi,\mu,\mu')].$$

$$(8.2\text{-}4)$$

Here η is the mean-free-path position of the unit source.

The scattering-component forms of these equations, i.e., the unscattered, single-scattered, and multiscattered, are not usually necessary since higher angular quadrature is easily utilized as the size of the discrete system is markedly reduced when energy dependence is eliminated. However, it is customary to break out the unscattered contribution and treat it separately. This will be done for the transmission function when appropriate.

It is noted that the nonlinear portion of these equations involving two integrals is coupled together by the scattering law, $f(\xi,\mu',\mu'') d\mu'$, and under the assumption of isotropic scattering a decoupling will result.

8.3. HOMOGENEOUS MEDIA WITH ISOTROPIC SCATTERING

In the restriction to a homogeneous media the properties of the medium become constants; thus the ξ-dependence can be dropped for these values and retained only in the functional unknowns. This restriction is for convenience only and in actual practice no further effort is required to solve for the non-homogeneous case since initial-valued systems are involved and properties of the medium can be easily changed at each integration step.

The restriction to isotropic scattering is a severe restriction that is utilized so that computational comparisons can be more easily obtained. In actual practice this is a poor assumption for many problems and at least P_1-scattering or linearly anisotropic scattering is necessary. These two cases are

$$f_0(\mu', \mu'')d\mu' = d\mu'/2 \qquad (8.3\text{-}1)$$

and

$$f_1(\mu', \mu'')d\mu' = (1/2)\left[1 + a\mu'\mu''\right]d\mu'. \qquad (8.3\text{-}2)$$

This latter equation follows from Eq. (4.3-8) for the special case of ϕ-angle averaged scattering as this case reduces to

$$\cos\theta_s = \mu_s = \cos\theta'\cos\theta'' = \mu'\mu''. \qquad (8.3\text{-}3)$$

The subscripts on the f-values in Eqs. (8.3-1, 2) represent the order of the Legendre expansion that is utilized for the scattering angle. In this section only the f_0 result will be employed.

In utilizing isotropic scattering, it is also common to redefine the input directions, μ_0 and μ', so as to obtain positive values and eliminate the absolute value signs. Thus using the input angles with respect to the input vector normals to the surface, these values have the same value but are now positive. In addition the limits on the integrals become zero to one. Formally this represents a $180°$ transformation of these input angles.

Employing isotropic scattering in a homogeneous slab, the results are

$$\frac{d}{d\xi}\hat{R}(\xi, \mu, \mu_0) = -\left[1/\mu + 1/\mu_0\right]\hat{R}(\xi, \mu, \mu_0) + (c/2)\Gamma(\xi, \mu)\Gamma(\xi, \mu_0), \qquad (8.3\text{-}4)$$

$$\Gamma(\xi, s) = 1 + \int_0^1 d\mu'\,\hat{R}(\xi, s, \mu')/\mu', \qquad (8.3\text{-}5)$$

$$\hat{R}(\xi, \mu, \mu_0) = \mu_0\,R(\xi, \mu, \mu_0), \qquad (8.3\text{-}6)$$

$$\hat{R}(\xi, \mu, \mu_0) = \hat{R}(\xi, \mu_0, \mu), \qquad (8.3\text{-}7)$$

$$T_0(\xi, \mu_0) = \exp(-\xi/\mu_0), \qquad (8.3\text{-}8)$$

$$\frac{d}{d\xi}\hat{T}_s(\xi, \mu, \mu_0) = -\hat{T}_s(\xi, \mu, \mu_0)/\mu_0 + (c/2)\Gamma(\xi, \mu_0)\,[T_0(\xi, \mu)$$

$$+\int_0^1 d\mu'\,\hat{T}_s(\xi, \mu, \mu')/\mu'], \qquad (8.3\text{-}9)$$

$$T(\xi, \mu, \mu_0) = T_0(\xi, \mu_0) + T_s(\xi, \mu, \mu_0), \qquad (8.3\text{-}10)$$

$$\hat{T}_s(\xi, \mu, \mu_0) = \mu_0\,T_s(\xi, \mu, \mu_0), \qquad (8.3\text{-}11)$$

$$\hat{T}_s(\xi, \mu, \mu_0) = \hat{T}_s(\xi, \mu_0, \mu), \qquad (8.3\text{-}12)$$

$$\frac{d}{d\xi}E_R(\xi, \mu) = -E_R(\xi, \mu)/\mu + S(\xi, \mu) + \int_0^1 d\mu'\,S(\xi, \mu')\,\hat{R}(\xi, \mu, \mu')/\mu'$$

$$+ (c/2)\Gamma(\xi, \mu)\int_0^1 d\mu''\,E_R(\xi, \mu'')/\mu'', \qquad (8.3\text{-}13)$$

and

$$\frac{d}{d\xi} G_R(\xi,\mu,\eta,\mu_0) = - G_R(\xi,\mu,\eta,\mu_0)/\mu + (c/2)\Gamma(\xi,\mu)$$
$$\cdot \int_0^1 d\mu'' G_R(\xi,\mu'',\eta,\mu_0)/\mu''. \qquad (8.3\text{-}14)$$

Here for the transmission function the subscript s refers to the scattered component. It is noted that the transformations to the $\hat{R}$ and $\hat{T}$ variables allow a symmetric form for the equations and thus in the computational phase the order of the system can be reduced. On the other hand, if the full system is utilized an accuracy check on the calculations can be easily deduced. Physically these symmetric functions represent the result for an isotropic input to the slab and thus the symmetric answer is immediately expected when isotropic scattering is allowed.

The initial conditions for these equations are

$$\hat{R}(0,\mu,\mu_0) = 0, \qquad (8.3\text{-}15)$$

$$\hat{T}_s(0,\mu,\mu_0) = 0, \qquad (8.3\text{-}16)$$

$$E_R(0,\mu) = B(\mu), \qquad (8.3\text{-}17)$$

$$G_R^-(\eta,\mu,\eta,\mu_0) = R(\eta,\mu,\mu_0), \qquad (8.3\text{-}18)$$

and

$$G_R^+(\eta,\mu,\eta,\mu_0) = 0. \qquad (8.3\text{-}19)$$

Here the plus and minus nomenclatures of Chapter 7 are utilized to indicate an outwardly directed or inwardly directed angle, respectively, for the Green's function unit sources. In addition, a general boundary source has been indicated on the escape function. In the normal escape probability calculations, this $B(\mu)$ is zero but for the combined result of transmission through a slab with an internal source this will have an angular value.

It is noted that in the further restricted case of an internal source that is isotropic, i.e., independent of angle, then the escape function equation has the further reduced form

$$\frac{d}{d\xi} E_R(\xi,\mu) = - E_R(\xi,\mu)/\mu + \Gamma(\xi,\mu) [S(\xi) + (c/2)\int_0^1 d\mu'' E_R(\xi,\mu'')/\mu''].$$

$$(8.3\text{-}20)$$

8.4. THE CALCULATIONAL FORMS

In order to calculate the desired functions, some discrete form of the equations is necessary. Following the usual angular discrete notation of previous sections let

$$R_{ij}(\xi) = \hat{R}(\xi, \mu_i, \mu_j), \qquad (8.4\text{-}1)$$

$$T_{ij}(\xi) = \hat{T}_s(\xi, \mu_i, \mu_j), \qquad (8.4\text{-}2)$$

$$E_{Ri}(\xi) = E_R(\xi, \mu_i), \qquad (8.4\text{-}3)$$

and

$$G_{Rij}(\xi, \eta) = G_R(\xi, \mu_i, \eta, \mu_j). \qquad (8.4\text{-}4)$$

Thus the discrete versions become

$$\frac{d}{d\xi} R_{ij}(\xi) = - \left[1/\mu_i + 1/\mu_j \right] R_{ij}(\xi) + c/2 \, \Gamma_i(\xi)\Gamma_j(\xi), \qquad (8.4\text{-}5)$$

$$\Gamma_u(\xi) = 1 + \sum_{s=1}^{L} w_s R_{us}(\xi)/\mu_s, \qquad (8.4\text{-}6)$$

$$T_{0j}(\xi) = \exp\left(- \xi/\mu_j\right), \qquad (8.4\text{-}7)$$

$$\frac{d}{d\xi} T_{ij}(\xi) = - T_{ij}(\xi)/\mu_j + (c/2)\Gamma_j(\xi) \left[T_{0i}(\xi) + \sum_{s=1}^{L} w_s T_{is}(\xi)/\mu_s \right], \qquad (8.4\text{-}8)$$

$$\frac{d}{d\xi} E_{Ri}(\xi) = - E_{Ri}(\xi)/\mu_i + S_i(\xi) + \sum_{s=1}^{L} w_s S_s(\xi) R_{ij}(\xi)/\mu_s$$

$$+ (c/2)\Gamma_i(\xi) \sum_{u=1}^{L} w_u E_{Ru}(\xi)/\mu_u, \qquad (8.4\text{-}9)$$

and

$$\frac{d}{d\xi} G_{Rij}(\xi, \eta) = - G_{Rij}(\xi, \eta)/\mu_i + (c/2)\Gamma_i(\xi) \sum_{u=1}^{L} w_u G_{Ruj}(\xi, \eta)/\mu_u.$$

$$(8.4\text{-}10)$$

The boundary conditions in the discrete form are obvious from Eq. (8.3-15) through Eq. (8.3-19) and will not be written here. The numerical considerations are explained in the appendix in order to solve these systems of equations.

8.5. THE "BLACKNESS COEFFICIENT" PROBLEM

One typical problem that shows the advantages of these invariant imbedding equations for the reflection and transmission is the "Blackness Coefficient" problem. The blackness of a medium is defined as the probability that a particle entering it will be absorbed. The standard blackness coefficient problem utilizes an isotropic input so that it becomes

$$\beta(c,\xi) = 1 - 2\int_0^1 d\mu \int_0^1 d\mu_0 \left[\hat{R}(\xi,\mu,\mu_0) + \mu_0 T_0(\xi,\mu_0) + \hat{T}_s(\xi,\mu,\mu_0)\right]$$

$$= 1 - 2E_3(\xi) - 2\int_0^1 d\mu \int_0^1 d\mu_0 \left[\hat{R}(\xi,\mu,\mu_0) + \hat{T}_s(\xi,\mu,\mu_0)\right], \qquad (8.5\text{-}1)$$

where the E_n-functions (see Abramowitz (1964)) are defined by

$$E_n(\xi) = \int_0^1 d\mu\, \mu^{n-2} \exp(-\xi/\mu). \qquad (8.5\text{-}2)$$

Here the implied dependence upon the c-value has been noted as a functional dependence. In practice the numerical quadrature order used is adequate to produce high accuracy so that the T_0-function can be left under the integral sign. This is especially convenient if a subroutine for E_n-functions does not happen to be available. Equation (8.5-1) does indicate that the result for a black slab is

$$\beta(0,\xi) = 1 - 2E_3(\xi) \qquad (8.5\text{-}3)$$

which is an important result in its own right. Thus the calculational form is

$$\beta(c,\xi) = 1 - 2\sum_{i=1}^L w_i \sum_{j=1}^L w_j \left[R_{ij}(\xi) + \mu_j T_{0j}(\xi) + T_{ij}(\xi)\right]. \qquad (8.5\text{-}4)$$

Table 2 shows the reflection function results for $\xi = 1$, $c = 0.5$ and $L = 7$, while Table 3 presents the transmission results. Table 4 gives a summary of some blackness calculations for a range of scattering parameters and slab thicknesses.

TABLE 2

Nonsymmetric Reflection Function Values[a] for Isotropic Scattering

$R_{ij}(\xi)/\mu_j$ ($L = 7$, $c = 0.5$, $\xi = 1.0$ mfp)

Output Angle, μ_i	Input angle, μ_j						
	0.025446	0.129234	0.297077	0.500000	0.702926	0.870766	0.974554
0.025446	0.131638	0.045803	0.023079	0.014648	0.010782	0.008863	0.007986
0.129234	0.232625	0.147202	0.093693	0.065538	0.050490	0.042452	0.038652
0.297077	0.269439	0.215376	0.161672	0.123553	0.099669	0.085833	0.079025
0.500000	0.287829	0.253563	0.207948	0.167472	0.138921	0.121396	0.112528
0.702926	0.297841	0.274624	0.235829	0.195301	0.164413	0.144776	0.134678
0.870766	0.303276	0.286039	0.251587	0.211414	0.179345	0.158550	0.147761
0.974554	0.305870	0.291473	0.259240	0.219330	0.186722	0.165373	0.154249

[a] Calculations are performed with a 12-digit word length using Runge–Kutta–Gill numerical integration with a step size of 0.02.

TABLE 3

Nonsymmetric Scattered and Direct Transmission Values[a] for Isotropic Scattering

$$T_{ij}(\xi)/\mu_j; T_{0j}(\xi) \quad (L = 7, c = 0.5, \xi = 1.0 \text{ mfp})$$

Output Angle, μ_i	Input angle, μ_j						
	0.025446	0.129234	0.297077	0.500000	0.702926	0.870766	0.974554
0.025446	0.000685	0.000881	0.001961	0.003085	0.003499	0.003555	0.003525
0.129234	0.004476	0.006349	0.013968	0.019769	0.021242	0.021023	0.020602
0.297077	0.022892	0.032109	0.050156	0.058381	0.057797	0.055095	0.053115
0.500000	0.060620	0.076485	0.098259	0.103096	0.097509	0.090980	0.086886
0.702926	0.096647	0.115537	0.136756	0.137082	0.126947	0.117256	0.111475
0.870766	0.121649	0.141651	0.161491	0.158445	0.145254	0.133507	0.126646
0.974554	0.135000	0.155359	0.174242	0.169350	0.154552	0.141741	0.134324
Direct Transmission	0.000000	0.000436	0.034523	0.135335	0.241079	0.317139	0.358398

[a]Calculations are performed with a 12-digit word length using Runge–Kutta–Gill numerical integration with a step size of 0.02.

TABLE 4

Blackness Coefficients[a] for Slabs with Isotropic Scattering ($L = 7$)

c-value	Thickness, ξ (mfp)				
	0.2	0.4	0.6	0.8	1.0
0.0	0.296108	0.485427	0.616899	0.711352	0.780616
0.1	0.273567	0.454725	0.583343	0.677405	0.747394
0.2	0.259826	0.421389	0.546255	0.639371	0.709793
0.3	0.224751	0.385107	0.505020	0.596415	0.666828
0.4	0.198226	0.345463	0.458882	0.547474	0.617194
0.5	0.170119	0.301959	0.406887	0.491156	0.559123
0.6	0.140284	0.253995	0.347818	0.425593	0.490149
0.7	0.108554	0.200838	0.280093	0.348228	0.406733
0.8	0.074745	0.141585	0.201618	0.255456	0.303602
0.9	0.038641	0.075113	0.109561	0.142028	0.172542
1.0	0.000000	0.000000	0.000000	0.000000	0.000000

[a]Calculations are performed with a 12-digit word length using Runge–Kutta–Gill numerical integration with a step size of 0.02.

8.6. THE ESCAPE PROBABILITY PROBLEM

The natural example to employ utilizing the escape function is the escape probability from a homogeneous, isotropically scattering slab containing various sources. Thus the escape probability becomes

$$P(c,\xi) = \frac{\int_0^1 d\mu \, [E_R(\xi,\mu) + E_L(\xi,\mu)]}{\int_0^\xi d\xi \int_{-1}^{+1} d\mu \, S(\xi,\mu)} \tag{8.6-1}$$

and represents the ratio of the total leakage compared to the total source strength. In the simplest case a constant, isotropic source is employed so that Eq. (8.6-1) reduces to

$$P(c,\xi) = \frac{2}{\xi} \int_0^1 d\mu \, E_R(\xi,\mu) \tag{8.6-2}$$

since for this symmetric case the E_L and E_R functions will be identical. In discrete form this is

$$P(c,\xi) = \frac{2}{\xi} \sum_{i=1}^{L} w_i \, E_{Ri}(\xi). \tag{8.6-3}$$

Naturally, the $B(\mu)$ of Eq. (8.3-17) is zero for these escape function calculations. Table 5 shows some calculated results for a range of c-values for a slab of one mean-free-path thickness. Also shown are the results of utilizing a variational

TABLE 5

Total Slab Escape Probabilities, Isotropic and Uniform Source (ξ = 1.0 mfp)

	Invariant Imbedding			Variational		
c-value	$L = 7$ $h = 0.02$	$L = 7$ $h = 0.01$	$L = 11$ $h = 0.01$	$\phi_0 = a$	$\phi_0 = a + bx^2$	$\phi_0 = a \cosh \kappa x$
0.0	0.390307	0.390307	0.390308	0.390308	0.390308	0.390308
0.1	0.415219	–	0.415221	0.415650	0.415229	0.415227
0.2	0.443621	–	0.443623	0.444511	0.443637	0.443636
0.3	0.476306	–	0.476308	0.477679	0.476328	0.476326
0.4	0.514329	–	0.514331	0.516196	0.514356	0.514354
0.5	0.559123	0.559123	0.559126	0.561470	0.559154	0.559151
0.6	0.612686	–	0.612689	0.615448	0.612719	0.612716
0.7	0.677888	–	0.677891	0.680910	0.677920	0.677917
0.8	0.759006	–	0.759008	0.761954	0.759033	0.759032
0.9	0.862709	–	0.862710	0.864896	0.862727	0.862726
1.0	1.000000	1.000000	1.000000	1.000000	1.000000	1.000000

calculation of the escape probability as per the method of Francis (1958). It is noted that this variational procedure predicts an upper boundary on the true result and thus the increasingly more accurate trial functions that are used approach the true answer from above. These values allow comparison with the invariant imbedding results for various quadratures as well as two different numerical integration increments. Thus the total accuracy of the table for the $L = 11$ case is expected to be better than five places. Table 6 is a similar calcu-

TABLE 6

Total Slab Escape Probabilities, Isotropic and Uniform Source $(c = 0.5)$

ξ (mfp)	Invariant Imbedding			Variational		
	$L = 7$ $h = 0.02$	$L = 7$ $h = 0.01$	$L = 11$ $h = 0.01$	$\phi_0 = a$	$\phi_0 = a + bx^2$	$\phi_0 = a \cosh \kappa x$
0.0	1.000000	1.000000	1.000000	1.000000	1.000000	1.000000
0.1	0.911022	0.911025	0.911298	0.911319	0.911312	0.911295
0.2	0.850594	0.850594	0.850660	0.850755	0.850660	0.850655
0.3	0.799710	0.799710	0.799665	0.799900	0.799675	0.799670
0.4	0.754897	0.754897	0.754863	0.755278	0.754867	0.754866
0.5	0.714672	0.714672	0.715662	0.715306	0.714669	0.714668
0.6	0.678145	0.678145	0.678150	0.679068	0.678159	0.678158
0.7	0.644715	0.644715	0.644724	0.645955	0.644737	0.644736
0.8	0.613945	0.613945	0.613952	0.615529	0.613970	0.613969
0.9	0.585502	0.585502	0.585507	0.587456	0.585529	0.585528
1.0	0.559123	0.559123	0.559126	0.561470	0.559154	0.559151

lation for a constant c-value of one-half and varying slab thicknesses. It shows that only for very thin slabs, $\xi < 0.3$, does the $L = 11$ results exceed the best variational values so that five place accuracy is all that is attained. It is noted that the invariant imbedding results in Table 6 for a fixed L-value are obtained in the same calculation whereas each variational result is a separate problem.

In this restricted constant isotropic source problem Warming and Heaslet (1966) have shown the following relationship between the escape probability and the blackness coefficient

$$\beta(c,\xi) = 2\xi(1 - c) P(c,\xi). \tag{8.6-4}$$

Thus in this restricted problem the evaluation of the escape function and the reflection function will produce sufficient information to evaluate the blackness coefficient also. In this example as with other examples, when an integrated result is desired the use of the escape function is preferred over that of the transmission function.

8.7. THE CRITICAL SLAB PROBLEM

A phenomenon that is unique with the transport of neutrons is the critical reactor whereby a medium becomes multiplying because of the presence of fissionable material. This situation can be modeled in a first approximation by a monoenergetic thermal group of neutrons since the fission cross section of most fuel materials is high for thermal neutrons. In this case the c-value is greater than unity and represents the mean number of secondaries per collision, i.e.,

$$c = [\nu \Sigma_f + \Sigma_s]/\Sigma, \qquad (8.7\text{-}1)$$

where ν is the mean number of neutrons per fission.

Equation (8.7-1) is valid for the assumption of isotropic scattering. If the scattering is anisotropic, then the following result is employed

$$cf(\mu',\mu'')d\mu' = c_f \frac{d\mu'}{2} + c_s f(\mu',\mu'')d\mu', \qquad (8.7\text{-}2)$$

where

$$c_f = \nu \Sigma_f / \Sigma, \qquad (8.7\text{-}3)$$

$$c_s = \Sigma_s / \Sigma, \qquad (8.7\text{-}4)$$

and

$$\Sigma = \Sigma_f + \Sigma_c + \Sigma_s. \qquad (8.7\text{-}5)$$

Here Σ_c is the nonfission capture cross section. It is noted that the fission process is considered isotropic in Eq. (8.7-2). In this section Eq. (8.7-1) is assumed to apply.

For a medium with a fixed c-value greater than unity, a size of the medium will be reached where the system becomes critical, i.e., self-substaining. In this case the reflection function for the medium which represents the reflection per unit input will grow without bounds and become infinite. Thus

$$R(\xi_c,\mu,\mu_0) = \infty \qquad (8.7\text{-}6)$$

is the condition indicating a critical size, ξ_c.

In the numerical solution of the reflection function it is difficult to determine accurately when Eq. (8.7-6) is satisfied. To overcome this problem, a transformation is made as follows:

$$\Lambda(\xi,\mu,\mu_0) = \frac{\hat{R}(\xi,\mu,\mu_0) - 1}{\hat{R}(\xi,\mu,\mu_0) + 1}. \qquad (8.7\text{-}7)$$

Thus the initial condition is

$$\Lambda(0,\mu,\mu_0) = -1 \qquad (8.7\text{-}8)$$

while the critical condition is

$$\Lambda(\xi_c,\mu,\mu_0) = +1. \tag{8.7-9}$$

Utilizing Eq. (8.7-7) in Eq. (8.3-4) produces the differential equation

$$\frac{d}{d\xi} \Lambda(\xi,\mu,\mu_0) = - (1/2) [1/\mu + 1/\mu_0] [1 - \Lambda^2(\xi,\mu,\mu_0)]$$

$$+ (c/4) [1 - \Lambda(\xi,\mu,\mu_0)]^2 \Gamma(\xi,\mu) \Gamma(\xi,\mu_0), \tag{8.7-10}$$

where now

$$\Gamma(\xi,s) = 1 + \int_0^1 d\mu'/\mu' [1 + \Lambda(\xi,s,\mu')] [1 - \Lambda(\xi,s,\mu')]^{-1}. \tag{8.7-11}$$

The numerical form for this is naturally

$$\frac{d}{d\xi} \Lambda_{ij}(\xi) = - (1/2) [1/\mu_i + 1/\mu_j] [1 - \Lambda_{ij}^2(\xi)]$$

$$+ (c/4) [1 - \Lambda_{ij}(\xi)]^2 \Gamma_i(\xi) \Gamma_j(\xi) \tag{8.7-12}$$

and

$$\Gamma_s(\xi) = 1 + \sum_{u=1}^{L} w_u/\mu_u [1 + \Lambda_{us}(\xi)] [1 - \Lambda_{us}(\xi)]^{-1}. \tag{8.7-13}$$

This functional is stable as the critical size is approached and can be made less subject to numerical roundoff by using the integrated form

$$\Lambda(\xi) = \int_0^1 d\mu \int_0^1 d\mu_0 \Lambda(\xi,\mu,\mu_0) \tag{8.7-14}$$

Table 7 gives some results with various quadratures and also shows some further comparisons with standard S_N-theory calculations, Carlson (1958), and a variational method, Francis (1958).

TABLE 7

Critical Size of a Bare Multiplying Slab ($c = 2.0$)

Invariant Imbedding		S_N Method		Variational Method	
Order of calculations L	Critical thickness ξ_c(mfp)	Order of calculations L	Critical thickness ξ_c(mfp)	Trial function	Critical thickness ξ_c(mfp)
3	0.62043	2	0.8194	a_1	0.63644
5	0.62200	4	0.6526	$a_1 + a_2 x^2$	0.62208
7	0.62195	8	0.6274		
11	0.62193	16	0.6232		

It should be noted that in this formulation a nonconstant c-value can be handled in a straightforward manner and thus, for instance, a series of layered slabs offer no additional complications.

A second procedure that can be utilized to obtain critical conditions permits the calculation of the critical c-value necessary for a particular sized slab. Since in the invariant imbedding method the calculations proceed by integrating from a zero-sized slab on to larger and larger sizes, an estimate of the critical c-value can be made at any step.

The key to this procedure is to expand the reflection function into a power series in the c-value. This in essence creates an order-of-scattering expansion to the reflection function. Thus letting

$$\hat{R}(\xi,\mu,\mu_0) = \sum_{n=1}^{\infty} c^n \hat{R}_n(\xi,\mu,\mu_0), \tag{8.7-15}$$

where the $\hat{R}_0$ term is zero and is excluded from the sum. Substituting this expansion into Eq. (8.3-4)

$$\sum_{n=1}^{\infty} c^n \frac{d}{d\xi} \hat{R}_n(\xi,\mu,\mu_0) = - [1/\mu + 1/\mu_0] \sum_{n=1}^{\infty} c^n \hat{R}_n(\xi,\mu,\mu_0)$$

$$+ (1/2) \{ c + \sum_{n=1}^{\infty} c^{n+1} [\Gamma_n(\xi,\mu) + \Gamma_n(\xi,\mu_0)]$$

$$+ \sum_{n=1}^{\infty} \sum_{m=1}^{\infty} c^{n+m+1} \Gamma_n(\xi,\mu)\Gamma_m(\xi,\mu_0) \}, \tag{8.7-16}$$

where

$$\Gamma_n(\xi,s) = \int_0^1 d\mu' \, \hat{R}_n(\xi,s,\mu')/\mu'. \tag{8.7-17}$$

Equating powers of c produces the following set of equations

$$\frac{d}{d\xi} \hat{R}_1(\xi,\mu,\mu_0) = - [1/\mu + 1/\mu_0] \hat{R}_1(\xi,\mu,\mu_0) + 1/2, \tag{8.7-18}$$

$$\frac{d}{d\xi} \hat{R}_2(\xi,\mu,\mu_0) = - [1/\mu + 1/\mu_0] \hat{R}_2(\xi,\mu,\mu_0) + 1/2 [\Gamma_1(\xi,\mu) + \Gamma_1(\xi,\mu_0)],$$

$$\tag{8.7-19}$$

$$\cdot$$
$$\cdot$$
$$\cdot$$

$$\frac{d}{d\xi} \hat{R}_p(\xi,\mu,\mu_0) = - [1/\mu + 1/\mu_0] \hat{R}_p(\xi,\mu,\mu_0) + 1/2 \{ \Gamma_{p-1}(\xi,\mu)$$

$$+ \Gamma_{p-1}(\xi,\mu_0) + \sum_{q=1}^{p-2} \Gamma_q(\xi,\mu) \, \Gamma_{p-1-q}(\xi,\mu_0) \} \quad (p > 2). \tag{8.7-20}$$

The general solution to this system can be represented as

$$\hat{R}_p(\xi,\mu,\mu_0) = \exp(-\tau\xi) \int_0^\xi S_p(z,\mu,\mu_0) \exp(\tau z) dz, \qquad (8.7\text{-}21)$$

where

$$S_p(z,\mu,\mu_0) = (1/2) \{\delta_{p1} + H(p-1) [\Gamma_{p-1}(z,\mu) + \Gamma_{p-1}(z,\mu_0)]$$

$$+ H(p-2) \sum_{q=1}^{p-2} \Gamma_q(z,\mu) \Gamma_{p-1-q}(z,\mu_0)\} \qquad (8.7\text{-}22)$$

and the Heaviside step function is

$$H(p) = \begin{cases} 0 & p \leqslant 0 \\ 1 & p > 0. \end{cases} \qquad (8.7\text{-}23)$$

The τ-function is

$$\tau = 1/\mu + 1/\mu_0. \qquad (8.7\text{-}24)$$

It is noted that the source term is always a function of the previously found lower orders of scattering. Naturally, the standard numerical solution of Eq. (8.7-19) and Eq. (8.7-20) can be performed as a set of coupled differential equations. Equation (8.7-18) has the analytical solution

$$\hat{R}_1(\xi,\mu,\mu_0) = \frac{1}{2\tau} [1 - \exp(-\tau\xi)]. \qquad (8.7\text{-}25)$$

The usage of these orders-of-scattering solutions to estimate the critical c-value is based upon the following:

$$R(\xi_c,\mu,\mu_0) = \infty = \sum_{n=1}^\infty c^n \hat{R}_n(\xi_c,\mu,\mu_0). \qquad (8.7\text{-}26)$$

Thus the series is diverging. Comparing two successive members of the series in the usual ratio test then

$$\frac{c^{p+1} \hat{R}_{p+1}(\xi_c,\mu,\mu_0)}{c^p \hat{R}_p(\xi_c,\mu,\mu_0)} \geqslant 1 \qquad (8.7\text{-}27)$$

since the $\hat{R}_p$-values are all positive. Using the equal sign and letting c_c be a critical c-value for a corresponding ξ-value

$$c_c = \frac{\hat{R}_p(\xi,\mu,\mu_0)}{\hat{R}_{p+1}(\xi,\mu,\mu_0)} \qquad (8.7\text{-}28)$$

Equation (8.7-28) represents an approximation to the critical c-values; however this estimate would be expected to approach a uniform value as the order of the scattering, p, increases. It is found that when sufficient scattering terms are included to allow Eq. (8.7-15) to produce an accurate result for the reflection function, the value predicted by Eq. (8.7-28) will also be accurate. In practice to overcome roundoff problems an integrated version is utilized, thus

$$c_c = \frac{\hat{R}_p(\xi)}{\hat{R}_{p+1}(\xi)},\tag{8.7-29}$$

where

$$\hat{R}_p(\xi) = \int_0^1 d\mu \int_0^1 d\mu_0 \ \hat{R}_p(\xi,\mu,\mu_0).$$

Table 8 gives numerical results obtained by using $p = 14$ in Eq. (8.7-25) and compares them with the results of Carlvik (1967) obtained by using an iterative integral formulation. Naturally Eq. (8.7-19) and Eq. (8.7-20) are employed as a discrete formulation and are solved as a simultaneous set of differential equations as follows

$$\frac{d}{d\xi}\hat{R}_{pij}(\xi) = -\tau_{ij}\hat{R}_{pij}(\xi) + 1/2 \ \{H(p-1) \ [\Gamma_{p-1,i}(\xi) + \Gamma_{p-1,j}(\xi)]$$

$$+ H(p-2)\sum_{q=1}^{p-2} \Gamma_{qi}(\xi) \ \Gamma_{p-1-q,j}(\xi)\}$$

$$(p = 2, 3, \ldots, P; i = 1, 2, \ldots, L; j = i + 1, i + 1, i + 2, \ldots, L).\tag{8.7-30}$$

Here

$$\Gamma_{ps}(\xi) = \sum_{u=1}^{L} w_u \hat{R}_{psu}(\xi)/\mu_u\tag{8.7-31}$$

TABLE 8

Critical Multiplication Factor for Bare Slabs
(Scattering order is 15)

Slab thickness ξ(mfp)	c_c-value		
	$L = 5$	$L = 7$	Carlvik (1967)
0.1	6.1578	6.1060	6.1170
0.2	3.8142	3.8274	3.8303
0.3	2.9716	2.9788	2.9787
0.4	2.5202	2.5228	2.5225
0.5	2.2344	2.2351	2.2350
0.6	2.0360	2.0360	2.0360
0.7	1.8897	1.8895	1.8895
0.8	1.7772	1.7771	1.7771
0.9	1.6880	1.6879	1.6879
1.0	1.6155	1.6154	1.6154

and

$$\tau_{ij} = 1/\mu_i + 1/\mu_j. \qquad (8.7\text{-}32)$$

Equation (8.7-25) is employed for the single scattering result as

$$\hat{R}_{1\,ij}(\xi) = \frac{1}{2\tau_{ij}} \, [1 - \exp(-\tau_{ij}\xi)] . \qquad (8.7\text{-}33)$$

The high accuracy of the results in Table 8 is noted.

8.8. MULTIPLE SLAB PROBLEMS

The utilization of invariant imbedding methods for multiple slab problems poses no particular problems since in the formulation the cross sections and mean number of secondaries per collision can be functions of position and thus can be changed numerically as the numerical integration proceeds from slab to slab. However, there are also matrix methods for multiple slab problems based upon the calculation of reflection and transmission values for each individual slab. For instance, the work of Aronson (1966) is based upon this procedure and is called the transfer matrix method.

This method can be explained briefly by noting the reflection from two adjacent slabs of mean free path thicknesses ξ_1 and ξ_2 and c-values c_1 and c_2, respectively, with slab number one on the left. The reflection functions are $R(\xi_\gamma,\mu,\mu_0)$ for $\gamma = 1, 2$ with a like nomenclature for the transmission function. Thus the reflection from both slabs is

$$R(\xi_1 + \xi_2,\mu,\mu_0)d\mu = R(\xi_2,\mu,\mu_0)d\mu + \int_0^1 T(\xi_2,\mu'',\mu_0)d\mu'$$
$$\cdot \int_0^1 R(\xi_1,\mu'',\mu')d\mu'' \, \{T(\xi_1,\mu,\mu'')d\mu$$
$$+ \int_0^1 \int_0^1 R(\xi_2,\mu''',\mu'')d\mu''' \, R(\xi_1,\mu'',\mu''')d\mu'' \, T(\xi,\mu,\mu'')d\mu$$
$$+ \cdots \}. \qquad (8.8\text{-}1)$$

This type of procedure leads naturally to a geometric series but generally is an unsatisfactory method for obtaining the desired results.

A better procedure is to determine the total right moving current at the slab boundaries and then utilize this as an input to the second slab in a transmission calculation.

Defining the current as follows:

$j^+(\mu)d\mu$ = the right moving partial current at the interface between the slabs in the direction $d\mu$ about μ.

$j^-(\mu)d\mu$ = the left moving partial current at the interface between the slabs in the direction $d\mu$ about μ.

Therefore, in this reflection problem with an input on the right side of the second slab

$$j^-(\mu)d\mu = T(\xi_2,\mu,\mu_0)d\mu + \int_0^1 j^+(\mu')d\mu' R(\xi_2,\mu,\mu')d\mu \qquad (8.8\text{-}2)$$

and

$$j^+(\mu)d\mu = \int_0^1 j^-(\mu'')d\mu'' R(\xi_1,\mu,\mu'')d\mu \qquad (8.8\text{-}3)$$

producing the integral equation

$$j^-(\mu) = T(\xi_2,\mu,\mu_0) + \int_0^1 d\mu' R(\xi_2,\mu,\mu') \int_0^1 d\mu'' j^-(\mu'') R(\xi_1,\mu',\mu''), \quad (8.8\text{-}4)$$

where it is understood that the equation is also a function of the input angle μ_0. It is noted that the same transmission and reflection functions apply to an input on either side of the slab in the homogeneous medium case.

Equation (8.8-4) represents an integral equation that can be solved by any of the standard methods. By expressing this in the discrete form

$$j_i^- = T_{ij}(\xi_2) + \sum_{s=1}^{L} w_s R_{is}(\xi_2) \sum_{u=1}^{L} w_u R_{su}(\xi_1)j_u^-. \qquad (8.8\text{-}5)$$

The equation can be solved by iteration utilizing a zero initial guess. On the other hand if Equation (8.8-5) is expressed in matrix form

$$J^- = T_{2j} + R_2 R_1 J^-, \qquad (8.8\text{-}6)$$

where J^- is the vector of j_i^- components and R is a square reflection array, containing the weighting functions, then

$$J^- = [I - R_2 R_1]^{-1} T_{2j}. \qquad (8.8\text{-}7)$$

There is an obvious difficulty with this procedure represented by Eq. (8.8-7) when computed results are desired. The matrix inversion in this equation is difficult to obtain once the order of the system begins to become large. In this monoenergetic case most systems are small enough so that this is not too great a problem; however for other cases the inversion is not practical by direct methods and if iterative procedures are used then Eq. (8.8-5) might as well be employed.

Finally to finish the problem

$$R_{ij}(\xi_1 + \xi_2) = \sum_{s=1}^{L} w_s j_{sj}^+ T_{is}(\xi_2) + R_{ij}(\xi_1)$$

$$= \sum_{s=1}^{L} w_s \sum_{u=1}^{L} w_u j_{uj}^- R_{su}(\xi_1) T_{is}(\xi_2) + R_{ij}(\xi_1), \qquad (8.8\text{-}8)$$

where the subscript j has been added to the partial currents to complete the problem. In matrix form this is

$$R_{1+2} = T_2 R_1 J^- + R_1$$

$$= T_2 R_1 [I - R_2 R_1]^{-1} T_2 + R_1, \qquad (8.8\text{-}9)$$

where now the weights are included in the transmission factors also where appropriate.

There are some advantages to this matrix formulation that may not be apparent. First let $\xi_1 = \xi_2$ and $c_1 = c_2$ so that the result in Eq. (8.8-9) represents the reflection from a slab of twice the thickness. This doubling formula can be useful in certain instances as

$$R(2\xi) = T(\xi) R(\xi) [I - R(\xi) R(\xi)]^{-1} T(\xi) + R(\xi). \qquad (8.8\text{-}10)$$

Again if the system is to be critical, the reflection from the double slab is infinite implying that

$$\det (I - R(\xi) R(\xi)) = 0. \qquad (8.8\text{-}11)$$

Thus calculating the critical half thickness is all that is necessary in this formulation.

In general for two different material slabs

$$\det (I - R_1 R_2) = 0 \qquad (8.8\text{-}12)$$

as thus a one-sided reflected multiplying core problem could be considered. However, the extension of these procedures to more than two slabs produces much more involved matrix equations.

These procedures essentially represent the transfer matrix method and other composite results can be found by similar matrix operations.

8.9. DISADVANTAGE FACTOR PROBLEM

In heterogeneous nuclear reactor theory one quantity of interest is the disadvantage factor when the reactor can be assumed to consist of repeating arrays of unit cells. A unit cell is a fuel assembly, coolant assembly, and moderator assembly that can be considered as an isolated unit. In many instances a model consisting of just fuel and moderator is adequate.

In slab geometry in water reactors, the coolant and the moderator are the same quantity and a two region unit cell is a good model. Here the center-lines of the fuel and moderator represent lines of symmetry and thus are the unit cell boundaries. The disadvantage factor is defined as the average flux in the moderator (water) as compared to the average flux in the fuel (uranium). Here the term flux implies a nondirectional indication of the thermal neutron intensity and becomes in terms of partial currents

$$\phi(z) = \int_0^1 d\mu \, j^+(z, \mu)/\mu + \int_{-1}^0 d\mu \, j^-(z, \mu)/|\mu|. \qquad (8.9\text{-}1)$$

The flux is the quantity that expresses the correct reaction rate defined by

$$\left.\begin{array}{l} \text{No. of interactions} \\ \text{per unit volume} \\ \text{per unit time} \end{array}\right\} = \Sigma\phi. \tag{8.9-2}$$

Thus the disadvantage factor, which is so named because it shows the "disadvantage" that the fuel has in respect to the moderator in capturing neutrons, becomes

$$d = \frac{\int_{V_2} \phi dV}{\int_{V_1} \phi dV}, \tag{8.9-3}$$

where V_2 is the moderator volume and V_1 is the fuel volume.

The unit cell model has a slowing down source of thermal neutrons in the moderator region and in this model this source is taken as constant and isotropic. The unit cell is isolated from any net neutron migration, i.e., the total net current is zero at the boundaries, or

$$J(0) = J(z_1 + z_2) = 0, \tag{8.9-4}$$

where

$$J(z) = \int_0^1 d\mu \, j^+(z,\mu) - \int_{-1}^0 d\mu \, j^-(z,\mu). \tag{8.9-5}$$

Since the total source strength is absorbed within the unit cell, then

$$\int_{z_1}^{z_1+z_2} S(z)dz = \int_0^{z_1} \phi_1(z) \Sigma_{a1} dz + \int_{z_1}^{z_1+z_2} \phi_2(z) \Sigma_{a2} dz \tag{8.9-6}$$

and

$$Sz_2 = \Sigma_{a1} z_1 \overline{\phi_1} + \Sigma_{a2} z_2 \overline{\phi_2}, \tag{8.9-7}$$

where the average value of the fluxes is indicated by a bar. On the other hand, since the fuel has no internal source, then the total absorption in the fuel region must come from a net current flow of neutrons into the fuel region from the moderator region. This net current at the boundary is

$$-J(z_1) = \int_0^1 d\mu \, j^+(z_1,\mu) - \int_{-1}^0 d\mu \, j^-(z_1,\mu)$$
$$= \Sigma_{a1} z_1 \overline{\phi_1}. \tag{8.9-8}$$

The negative sign is necessary since the net migration into the fuel represents a minus z-direction flow of neutrons. Using Eq. (8.9-7) and Eq. (8.9-8) produces

$$d = \frac{\overline{\phi_2}}{\overline{\phi_1}} = \frac{(1 - c_1)\xi_1}{(1 - c_2)\xi_2} \left\{ \frac{S\xi_2}{-J(\xi_1)} - 1 \right\}, \tag{8.9-9}$$

where the c-values have been utilized to replace the absorption cross sections and the ξ are now measured in mean free paths in the respective slabs.

Therefore, the key to the calculation of the disadvantage factor problem is determining the net total current at the boundary between the two regions. In practice the μ-values are taken as positive for both j^+ and j^- since it is known that j^- is in the negative ξ-direction. So in discrete form

$$-J(\xi_1) = -\sum_{i=1}^{L} w_i [j_i^+(\xi_1) - j_i^-(\xi_1)], \qquad (8.9\text{-}10)$$

where

$$j_i^+(\xi) = E_{Ri}(\xi) + \sum_{s=1}^{L} w_s\, j_s^-(\xi)\, R_{is}(\xi) \qquad (8.9\text{-}11)$$

and

$$j_i^-(\xi) = E_{Li}(\xi) + \sum_{u=1}^{L} w_u\, j_u^+(\xi)\, R_{iu}(\xi_1 + \xi_2 - \xi). \qquad (8.9\text{-}12)$$

Here it is noted that these are not the $\hat{R}$-functions that are normally calculated but that

$$R_{ij}(\xi) = \hat{R}_{ij}(\xi)/\mu_j. \qquad (8.9\text{-}13)$$

The solution of Eqs. (8.9-11) and (8.9-12) by iteration with an initial zero starting condition is straightforward. At the boundaries these equations reduce to

$$j_i^+(0) = j_i^-(0) \qquad (8.9\text{-}14)$$

and

$$j_i^+(\xi_1 + \xi_2) = j_i^-(\xi_1 + \xi_2). \qquad (8.9\text{-}15)$$

It is noted that the escape function as well as the reflection function represent cumulative values throughout both regions except at the boundary between the regions. Here individual slab values are used when Eq. (8.9-10) is being evaluated.

It is also noted that the escape functions include the effect of a boundary source being transmitted through the medium, thus the initial conditions become

$$E_{Ri}(0) = j_i^+(0) \qquad (8.9\text{-}16)$$

and

$$E_{Li}(\xi_1 + \xi_2) = j_i^-(\xi_1 + \xi_2), \qquad (8.9\text{-}17)$$

where the E_L function is integrated from $\xi_1 + \xi_2$ to zero. However, since neither of these values is known, an estimate is made of one of them and an iterative procedure is set up. In practice assuming $j_i^+(0)$ is zero is a satisfactory starting value for the outer-iteration scheme. It is noted that the reflection functions need only to be computed once for this iteration since they do not change; however, because of computer storage limitations, it is usually easier to recompute them as they are needed.

The source strength in the moderator does not affect the calculation of the disadvantage factor and may be set equal to unity for convenience. However, it should be noted that the source required in the escape function differential equation, Eq. (8.4-9), is for only positive or negative μ-values; thus for a unity isotropic source in the moderator the value of one-half is necessary in these equations.

Table 9 gives the results for the solution to a typical problem having the parameter values

$$c_1 = 0.56890 \quad c_2 = 0.99276$$

$$\xi_1 = 0.21051 \quad \xi_2 = 0.81120$$

where the results of each iteration are noted. The convergence is rather slow for this sequence and improved values can be obtained by noting that the iteration appears monotone; thus utilizing Richardson's extrapolation, see Froberg (1965),

TABLE 9

Disadvantage Factor Iteration for a Typical Unit Cell with Isotropic Scattering
(L = 3, Zero Initial Angular Vector)

Iteration No.	Net current, J	Disadvantage factor	Extrapolated net current, J_∞	Extrapolated disadvantage factor
1	−0.793948	0.33571	−	−
2	−0.775841	0.70429	−0.757734	1.09050
3	−0.765491	0.92280	−0.755141	1.14730
4	−0.759484	1.05235	−0.753478	1.18397
5	−0.755994	1.12857	−0.752504	1.20550
6	−0.753966	1.17318	−0.751939	1.21803
7	−0.752788	1.19922	−0.751609	1.22534
8	−0.752103	1.21439	−0.751418	1.22958
9	−0.751705	1.22321	−0.751307	1.23204
10	−0.751474	1.22834	−0.751243	1.23348
11	−0.751339	1.23133	−0.751205	1.23431
12	−0.751261	1.23306	−0.751183	1.23480
13	−0.751216	1.23407	−0.751171	1.23508
14	−0.751190	1.23466	−0.751163	1.23524
15	−0.751174	1.23500	−0.751159	1.23535
16	−0.751165	1.23520	−0.751156	1.23539
17	−0.751160	1.23531	−0.751155	1.23542

with an error reduction factor of two produces

$$J_\infty = 2J_m - J_{m-1},\tag{8.9-18}$$

where J_∞ is an estimate of the final iterated value and m is an iteration index. However, it is noted that Eq. (8.9-9) is quite sensitive to the value of $J(\xi_1)$ and this value must be obtained to a high degree of accuracy.

This problem can be generalized quite easily to the case of nonisotropic scattering or a nonisotropic source. In fact sources may be placed in both regions also. Table 10 shows some results for utilizing an isotropic moderator

TABLE 10

Disadvantage Factor Values for a Typical Unit Cell with Anisotropic Moderator Scattering

Quadrature order, L	Anisotropic factor	Disadvantage factor, d
3	$a = 0.0$	1.2353
3	$a = 0.1$	1.2329
3	$a = 0.2$	1.2304
3	$a = 1.0$	1.2110
5	$a = 1.0$	1.2073
3	hydrogen	1.1670
5	hydrogen	1.1480

source but a nonisotropic moderator scattering. The scattering is linearly anisotropic as per Eq. (8.3-2) with various a-values in one case while the free hydrogen atom scattering kernel, see Davison (1958), is used in the other case. This is

$$f_H(\mu, \mu') = \begin{cases} |\gamma_1| + \gamma_1 & \gamma_2 > 1 \\ 2\gamma_1 [1 - \pi^{-1} \cos^{-1} (\gamma_1/[1 - \gamma_2 + \gamma_1^2]^{1/2})] \\ \quad + (2/\pi)[1 - \gamma_2]^{1/2} & \gamma_2 < 1 \end{cases}\tag{8.9-19}$$

where

$$\gamma_1 = \mu\mu'\tag{8.9-20}$$

and

$$\gamma_2 = \mu^2 + (\mu')^2.\tag{8.9-21}$$

In these results the parameter values for the unit cell, except for the scattering, are the same as before. The severe effect of the anisotropic scattering is noted and indicates to some degree the reason for the statement earlier in this chapter that isotropic scattering is in many instances a poor model to employ.

8.10. THE SEMIINFINITE MEDIA PROBLEM

In the semiinfinite media case the derivative with respect to the slab size becomes zero and the reflection function reduces to

$$[1/\mu + 1/\mu_0] \, \hat{R}(\mu, \mu_0) = (c/2) \, H(\mu) \, H(\mu_0), \qquad (8.10\text{-}1)$$

where isotropic scattering has been utilized and the Γ-functions have been replaced by the symbol H. This change in nomenclature is convenient since now these are the H-functions of Chandrasekar (1950). Naturally, the transmission function is zero for a semiinfinite medium while the escape function becomes for a uniform isotropic source

$$1/\mu \, E_R(\mu) = H(\mu) \, [S + (c/2) \int_0^1 d\mu'' \, E_R(\mu'')/\mu''] \, . \qquad (8.10\text{-}2)$$

The H-functions can be obtained by the iterative solution of the integral equation

$$\frac{1}{H(\mu)} = 1 - \mu(c/2) \int_0^1 d\mu' \, \frac{H(\mu')}{\mu' + \mu} \, . \qquad (8.10\text{-}3)$$

This is solved in the discrete formulation as

$$H_i^{-1} = 1 - \mu_i(c/2) \sum_{s=1}^{L} w_s \, \frac{H_s}{\mu_s + \mu_i} \qquad (8.10\text{-}4)$$

and generally converges quite rapidly except for c-values near unity. Naturally in this semiinfinite case the c-value cannot exceed unity.

Chandrasekar (1950) has also listed this H-function for isotropic scattering.

The corresponding iterative equation for the escape function is

$$E_{Ri} = \mu_i H_i \, [S + (c/2) \sum_{u=1}^{L} w_u \, E_{Ru}/\mu_u] \, . \qquad (8.10\text{-}5)$$

Here since the semiinfinite medium is critical for a c-value of unity, only c-values less than unity are feasible.

For this semiinfinite case an interesting approximation, see Mingle (1970), can be found for the reflection function by assuming that the function is separable. Thus let

$$\hat{R}(\mu, \mu_0) \cong G(\mu) \, G(\mu_0) \qquad (8.10\text{-}6)$$

and then evaluate the moments of the G-function

$$G_P = \int_0^1 d\mu \, \mu^P \, G(\mu). \qquad (8.10\text{-}7)$$

Substitution of Eq. (8.10-6) into Eq. (8.10-1) and utilizing Eq. (8.10-7) produces a set of nonlinear equations for the minus one, zero, and plus one moments of the G-function. The solution of these equations and evaluation of the reflection function gives

$$\hat{R}(\mu,\mu_0) \cong \frac{2c\mu\,\mu_0}{[1 + 2\mu\sqrt{1 - c}]\,[1 + 2\mu_0\sqrt{1 - c}]}.\qquad(8.10\text{-}8)$$

The approximation sign is to emphasize that the basic separable feature is an assumption. This approximation is within a few percent of the true result over most of the (μ,μ_0) space and thus can be used when a definite closed form equation is desirable but high accuracy is not required. For instance in the preliminary stages of optimization calculations, Eq. (8.10-8) could be conveniently employed and thus the true result utilized for only the final few iterations.

8.11. REFERENCES TO COMPUTED RESULTS

Many of the references listed in the bibliography section contain computed results for this energy-independent case. For instance the papers of Mingle (1966, 1967) both contain several computed tables of various results. In addition many other authors have also computed quantities in their papers.

8.12. EXERCISES

1. Determine the uncollided contribution to the escape function of Eq. (8.3-13).
2. A dissipation function is sometimes defined as:

$L(x,\mu_0)$ = the probability that a particle entering a slab of thickness x in the direction μ_0 will be absorbed.

Determine a differential equation for this function.

3. Using the dissipation function show that

$$L(x,\mu_0) + \int_0^1 d\mu\, R(x,\mu,\mu_0) + \int_0^1 d\mu\, T(x,\mu,\mu_0) = 1.\qquad(8.12\text{-}1)$$

4. Express the "Blackness Coefficient" in terms of the dissipation function.
5. Derive Eq. (8.6-4) from basic definitions.
6. Using linearly anisotropic scattering, Eq. (8.3-2), in Eq. (8.7-2) determine a relationship between c_f, a, and c_s in order for a medium to be critical.
7. Determine the set of equations for the transmission order-of-scattering expression as

$$\hat{T}(\xi,\mu,\mu_0) = \sum_{n=0}^{\infty} c^n\, \hat{T}_n(\xi,\mu,\mu_0).\qquad(8.12\text{-}2)$$

8. Using Eq. (8.8-1) deduce the geometric series applicable to the expansion and show its summed value.

9. Deduce integral equations for the partial currents, j^+ and j^-, for the internal region of a slab $(0 \leqslant z \leqslant x)$ that has distributed boundary sources $B_0(\mu)$ and $B_x(\mu)$.

10. Determine the transfer matrix formulation for the reflection from three layered slabs.

11. Using Eq. (8.9-1) for the flux determine the flux corresponding to the current requirements of Exercise 9.

12. Deduce the disadvantage factor expression similar to Eq. (8.9-9) for the case of a source in the fuel of strength βS where S is the moderator source strength.

13. Show the relationship between Eq. (8.10-3) and Eq. (3.5-1) in terms of the two H-functions.

CHAPTER 9

EXTENSIONS TO MULTIPARTICLE THEORY

9.1. INTRODUCTION

In the invariant imbedding treatment in the preceding chapters the assumption has been implied that all particles are alike and that no differentiation could be made between them. In many instances several types of particles can be identified, such as neutrons and photons, photons and annihilation photons, and photons and electrons. In general these different particles have sufficiently different physical interaction processes occurring that it becomes convenient to treat each of them separately. However, in many cases a certain sequence of interactions by one type of particle will produce another type of particle, such as the capture of a neutron releasing gamma photons.

In this chapter the general theory of multiparticle invariant imbedding transport theory is introduced in the usual functional and discrete forms.

9.2. BASIC DEFINITIONS

Define the following:

$\Sigma^h(x, E)dy$ = the probability of an interaction occurring when a particle of type-h and energy E traverses a distance dy at the position x.

$c^{hh'}(x, E')$ = the probability in an interaction of a particle of type-h' with energy E' producing a particle of type-h at position x.

$f^{hh'}(x, E, \Omega, E', \Omega')dE\,d\Omega$ = the probability in an interaction producing a particle of type-h from a particle of type h' of transferring from energy E' and direction Ω' into the energy range dE about E and the direction $d\Omega$ about Ω at the position x.

$R^{hh'}(x, E, \Omega, E', \Omega')dE\,d\Omega$ = the reflected particle current from a slab of thickness x of particles of type-h in the energy range dE about E in the direction $d\Omega$ about Ω because of a unit input at x of particles of type-h' of energy E' and direction Ω'.

$T^{hh'}(x, E, \Omega, E', \Omega')dE\, d\Omega$ = the transmitted particle current of type-h from a slab of thickness x in the energy range dE about E in the direction $d\Omega$ about Ω because of a unit input at x of particles of type-h' of energy E' and direction Ω'.

In the above definitions it is noted that when h and h' are identical then these functions are the same as those previously defined. Again the concepts of reflection and transmission automatically determine the physical ranges for the input and output angles with all other combinations being zero.

9.3. ALBEDO DERIVATION

In order to show the effect of particle type on the derivations, the reflection function will be followed through and only the results shown for the other functions of interest.

In reference to Fig. 4 which shows the five interactions for the reflection function, the following modified terms for H separate particle types are

1. $[1 - \Sigma^{h_0}(x, E_0)dx/|\mu_0|]\, R^{hh_0}(x, E, \Omega, E_0, \Omega_0)dE\, d\Omega[1 - \Sigma^h(x, E)dx/\mu]$,

2. $\Sigma^{h_0}(x, E_0)dx/|\mu_0| c^{hh_0}(x, E_0)f^{hh_0}(x, E, \Omega, E_0, \Omega_0)dE\, d\Omega$,

3. $\displaystyle\sum_{h'=1}^{H}\int_0^\infty\int_0^{4\pi}\Sigma^{h_0}(x, E_0)dx/|\mu_0| c^{h'h_0}(x, E_0)f^{h'h_0}(x, E', \Omega', E_0, \Omega_0)$

   $\cdot\, dE'd\Omega'\, R^{hh'}(x, E, \Omega, E', \Omega')dE\, d\Omega[1 - \Sigma^h(x, E)dx/\mu]$

4. $\displaystyle\sum_{h'=1}^{H}\int_0^\infty\int_0^{4\pi}[1 - \Sigma^{h_0}(x, E_0)dx/|\mu_0|]\, R^{h''h_0}(x, E'', \Omega'', E_0, \Omega_0)dE''\, d\Omega''$

   $\cdot\, \Sigma^{h''}(x, E'')dx/\mu'' c^{hh''}(x, E'')f^{hh''}(x, E, \Omega, E'', \Omega'')dE\, d\Omega$

5. $\displaystyle\sum_{h'=1}^{H}\sum_{h''=1}^{H}\int_0^\infty\int_0^\infty\int_0^{4\pi}\int_0^{4\pi}[1 - \Sigma^{h_0}(x, E_0)dx/|\mu_0|]$

   $\cdot\, R^{h''h_0}(x, E'', \Omega'', E_0, \Omega_0)dE''\, d\Omega''\, \Sigma^{h''}(x, E'')dx/\mu''$

   $\cdot\, c^{h'h''}(x, E'')f^{h'h''}(x, E', \Omega', E'', \Omega'')dE'\, d\Omega'$

   $\cdot\, R^{hh'}(x, E, \Omega, E', \Omega')dE\, d\Omega[1 - \Sigma^h(x, E)dx/\mu]$.

It is to be noted that in many applications certain transitions of particle types are impossible, thus the quantity c^{hh_0} is zero which automatically makes the corresponding albedo also zero.

The resulting differential equation becomes

$$\frac{d}{dx} R^{hh_0}(x,E,\Omega,E_0,\Omega_0) = -\ [\Sigma^{h_0}(x,E_0)/|\mu_0| + \Sigma^h(x,E)/\mu]$$

$$\cdot\ R^{hh_0}(x,E,\Omega,E_0,\Omega_0) + \Sigma^{h_0}(x,E_0) c^{hh_0}(x,E_0)$$

$$\cdot\ f^{hh_0}(x,E,\Omega,E_0,\Omega_0)/|\mu_0| + \Sigma^{h_0}(x,E_0)/|\mu_0|\ \sum_{h'=1}^{H} c^{h'h_0}(x,E_0)$$

$$\cdot\ \int_0^\infty dE'\ \int_0^{4\pi} d\Omega'\ f^{h'h_0}(x,E',\Omega',E_0,\Omega_0) R^{hh'}(x,E,\Omega,E',\Omega')$$

$$+ \sum_{h''=1}^{H} \int_0^\infty dE''\ \Sigma^{h''}(x,E'') c^{hh''}(x,E'') \int_0^{4\pi} d\Omega''$$

$$\cdot\ R^{h''h_0}(x,E'',\Omega'',E_0,\Omega_0) f^{hh''}(x,E,\Omega,E'',\Omega'')/\mu''$$

$$+ \sum_{h''=1}^{H} \int_0^\infty dE''\ \Sigma^{h''}(x,E'') \int_0^{4\pi} d\Omega''\ R^{h''h_0}(x,E'',\Omega'',E_0,\Omega_0)/\mu''$$

$$\cdot\ \sum_{h'=1}^{H} c^{h'h''}(x,E'') \int_0^\infty dE'\ \int_0^{4\pi} d\Omega'\ f^{h'h''}(x,E',\Omega',E'',\Omega'')$$

$$\cdot\ R^{hh'}(x,E,\Omega,E',\Omega')\qquad (h,h_0 = 1, 2, \cdots, H)\qquad\qquad (9.3\text{-}1)$$

with the usual initial condition

$$R^{hh_0}(0, E, \Omega, E_0, \Omega_0) = 0.\qquad\qquad (9.3\text{-}2)$$

It is noted that the integrations extend over the full range whereas in actual fact the reflection functional having certain zero ranges will effectively limit the integrations to the physically possible values.

The transformations of Chapters 3 and 5 can be performed on this multiparticle reflection equation; however, they do not undergo any particular complications because of multiparticle effects. Therefore the discrete forms will be listed later on in the chapter and the details left for an exercise.

9.4. THE TRANSMISSION RESULT

The result for the multiparticle transmission function becomes

$$\frac{d}{dx} T^{hh_0}(x,E,\Omega,E_0,\Omega_0) = -\ \Sigma^{h_0}(x,E_0)/|\mu_0| T^{hh_0}(x,E,\Omega,E_0,\Omega_0)$$

$$+ \Sigma^{h_0}(x,E_0) \sum_{h'=1}^{H} c^{h'h_0}(x,E_0)/|\mu_0| |\int_0^\infty dE' \int_0^{4\pi} d\Omega$$

$$\cdot\ f^{h'h_0}(x,E',\Omega',E_0,\Omega_0) T^{hh'}(x,E,\Omega,E',\Omega')$$

$$+ \sum_{h''=1}^{H} \int_0^\infty dE'' \int_0^{4\pi} d\Omega'' R^{h''h_0}(x,E'',\Omega'',E_0,\Omega_0) \, \Sigma^{h''}(x,E'')$$

$$\cdot \sum_{h'=1}^{H} c^{h'h''}(x,E'')/\mu'' \int_0^\infty dE' \int_0^{4\pi} d\Omega' f^{h'h''}(x\ E',\Omega',E'',\Omega'')$$

$$\cdot T^{hh'}(x,E,\Omega,E',\Omega') \quad (h,h_0 = 1,2,\cdots,H). \tag{9.4-1}$$

The initial condition now is

$$T^{hh_0}(0,E,\Omega,E_0,\Omega_0) = \delta_{hh_0} \delta(E-E_0)\delta(\Omega-\Omega_0). \tag{9.4-2}$$

9.5. THE UNSCATTERED CONTRIBUTIONS

In multiparticle theory the concept of the unscattered component requires clarification. For the pure particle this is straightforward as previously defined, but for the created particle an unscattered contribution exists after the creation. Since the creation is an interaction, this second type of unscattered result is actually a special form of first scattering where a $c^{hh'}$ transformation has occurred with no other h-type scattering. Therefore, let

$$R^{hh_0}(x,E,\Omega,E_0,\Omega_0) = R_1^{hh_0}(x,E,\Omega,E_0,\Omega_0)[1-\delta_{hh_0}]$$

$$+ R_M^{hh_0}(x,E,\Omega,E_0,\Omega_0) \tag{9.5-1}$$

producing

$$\frac{d}{dx} R_1^{hh_0}(x,E,\Omega,E_0,\Omega_0) = -\ [\Sigma^{h_0}(x,E_0)/|\mu_0|+\Sigma^h(x,E)/\mu] R_1^{hh_0}(x,E,\Omega,E_0,\Omega_0)$$

$$+ \Sigma^{h_0}(x,E_0) c^{hh_0}(x,E_0) f^{hh_0}(x,E,\Omega,E_0,\Omega_0)/|\mu_0| + \Sigma^{h_0}(x,E_0)/|\mu_0|$$

$$\cdot \sum_{\substack{h'=1 \\ h'\neq h}}^{H} c^{h'h_0}(x,E_0) \int_0^\infty dE' \int_0^{4\pi} d\Omega' f^{h'h_0}(x,E',\Omega,E_0,\Omega_0)$$

$$\cdot R_1^{hh'}(x,E,\Omega,E',\Omega') + \sum_{\substack{h''=1 \\ h''\neq h}}^{H} \int_0^\infty dE'' \Sigma^{h''}(x,E'') \, c^{hh''}(x,E'')$$

$$\cdot \int_0^{4\pi} d\Omega'' R^{h''h_0}(x,E'',\Omega'',E_0,\Omega_0) f^{hh''}(x,E,\Omega,E'',\Omega'')/\mu''$$

$$+ \sum_{\substack{h''=1 \\ h''\neq h}}^{H} \int_0^\infty dE'' \Sigma^{h''}(x,E'') \int_0^{4\pi} d\Omega'' R^{h''h_0}(x,E'',\Omega'',E_0,\Omega_0)/\mu''$$

$$\cdot \sum_{\substack{h''=1 \\ h''\neq h}}^{H} c^{h'h''}(x,E'') \int_0^\infty dE' \int_0^{4\pi} d\Omega' f^{h'h''}(x,E',\Omega',E'',\Omega'')$$

$$\cdot R_1^{hh'}(x,E,\Omega,E',\Omega') \quad (h\neq h_0) \tag{9.5-2}$$

and

$$\frac{d}{dx} R_M^{hh_0}(x,E,\Omega,E_0,\Omega_0) = - [\Sigma^{h_0}(x,E_0)/|\mu_0| + \Sigma^h(x,E)/\mu]$$

$$\cdot R_M^{hh_0}(x,E,\Omega,E_0,\Omega_0) + \Sigma^{h_0}(x,E_0)/|\mu_0| \sum_{h'=1}^{H} \int_0^\infty dE'$$

$$\cdot \int_0^{4\pi} d\Omega' c^{h'h_0}(x,E_0) f^{h'h_0}(x,E',\Omega',E_0,\Omega_0) R_M^{hh'}(x,E,\Omega,E',\Omega')$$

$$+ \sum_{h''=1}^{H} \int_0^\infty dE'' \Sigma^{h''}(x,E'') c^{hh''}(x,E'') \int_0^{4\pi} d\Omega''$$

$$\cdot [R_1^{h''h_0}(x,E'',\Omega'',E_0,\Omega_0)\delta_{hh''} + R_M^{h''h_0}(x,E'',\Omega'',E_0,\Omega_0)]/\mu''$$

$$\cdot f^{hh''}(x,E,\Omega,E'',\Omega'')(1 - \delta_{h''h_0}) + \sum_{h'=1}^{H} \int_0^\infty dE'' \Sigma^{h''}(x,E''(x,E'')$$

$$\cdot \int_0^{4\pi} d\Omega'' R^{h''h_0}(x,E'',\Omega'',E_0,\Omega_0)/\mu'' \sum_{h'=1}^{H} c^{h'h''}(x,E'')$$

$$\cdot \int_0^\infty dE' \int_0^{4\pi} d\Omega' f^{h'h''}(x,E',\Omega',E'',\Omega'')[R_1^{hh'}(x,E,\Omega,E',\Omega')$$

$$\cdot \delta_{h''h} + R_M^{hh'}(x,E,\Omega,E',\Omega')] \qquad (h,h_0 = 1,2,\cdots,H). \qquad (9.5\text{-}3)$$

Again R with no subscript implies $R_1 + R_M$. The terms like R_1^{hh} are always zero by definition; however, the delta function term in Eq. (9.5-1) accentuates this fact, while the $h' \neq h$ and $h'' \neq h$ restrictions on the summations do likewise for Eq. (9.5-2). These equations all have an initial condition of zero.

For the transmission function two unscattered contributions can be recognized. The first is the common form and will have identical particle-type designations. The second type will be comparable to the previous reflection type and will represent an interaction producing the desired particle but with no other scatterings before transmission. Therefore, let

$$T^{hh_0}(x,E,\Omega,E_0,\Omega_0) = T_0^{hh_0}(x,E,\Omega,E_0,\Omega_0)\delta_{hh_0}\delta(E - E_0)\delta(\Omega - \Omega_0)$$

$$+ T_1^{hh_0}(x,E,\Omega,E_0,\Omega_0)(1 - \delta_{hh_0}) + T_M^{hh_0}(x,E,\Omega,E_0,\Omega_0) \qquad (9.5\text{-}4)$$

producing

$$\frac{d}{dx} T_0^{h_0h_0}(x,E_0,\Omega_0,E_0,\Omega_0) = - \Sigma^{h_0}(x,E_0)/|\mu_0| T_0^{h_0h_0}(x,E_0,\Omega_0,E_0,\Omega_0), \qquad (9.5\text{-}5)$$

$$\frac{d}{dx} T_1^{hh_0}(x,E,\Omega,E_0,\Omega_0) = - \Sigma^{h_0}(x,E_0)/|\mu_0| T_1^{hh_0}(x,E,\Omega,E_0,\Omega_0)$$

$$+ \Sigma^{h_0}(x,E_0) c^{hh_0}(x,E_0)/|\mu_0| f^{hh_0}(x,E,\Omega,E_0,\Omega_0) T_0^{hh}(x,E,\Omega,E,\Omega)$$

$$+ \Sigma^{h_0}(x,E_0) \sum_{\substack{h'=1 \\ h' \neq h}}^{H} c^{h'h_0}(x,E_0)/|\mu_0| \int_0^\infty dE' \int_0^{4\pi} d\Omega'$$

$$\cdot f^{h'h_0}(x,E',\Omega',E_0,\Omega_0) T_1^{hh'}(x,E,\Omega,E',\Omega') + \sum_{h''=1}^{H}$$

$$\cdot \int_0^\infty dE'' \int_0^{4\pi} d\Omega'' R^{h''h_0}(x,E'',\Omega'',E_0,\Omega_0) \Sigma^{h''}(x,E'')$$

$$\cdot [c^{hh''}(x,E'')/\mu'' f^{hh''}(x,E,\Omega,E'',\Omega'') T_0^{hh}(x,E,\Omega,E,\Omega)(1 - \delta_{hh''})$$

$$+ \sum_{\substack{h'=1 \\ h' \neq h}}^{H} c^{h'h''}(x,E'')/\mu'' \int_0^\infty dE' \int_0^{4\pi} d\Omega' f^{h'h''}(x,E',\Omega',E'',\Omega'')$$

$$\cdot T_1^{hh'}(x \; E,\Omega,E',\Omega') \qquad (h \neq h_0), \tag{9.5-6}$$

and

$$\frac{d}{dx} T_M^{hh_0}(x,E,\Omega,E_0,\Omega_0) = - \Sigma^{h_0}(x,E_0)/|\mu_0| T_M^{hh_0}(x,E,\Omega,E_0,\Omega_0)$$

$$+ \Sigma^{h_0}(x,E_0) \sum_{h'=1}^{H} c^{h'h_0}(x,E_0)/|\mu_0| \int_0^\infty dE' \int_0^{4\pi} d\Omega'$$

$$\cdot f^{h'h_0}(x,E',\Omega',E_0,\Omega_0) T_M^{hh'}(x,E,\Omega,E',\Omega') + \sum_{h''=1}^{H}$$

$$\cdot \int_0^\infty dE'' \int_0^{4\pi} d\Omega'' R^{h''h_0}(x,E'',\Omega'',E_0,\Omega_0) \Sigma^{h''}(x,E'')$$

$$\cdot [c^{hh}(x,E'')/\mu'' f^{hh}(x,E,\Omega,E'',\Omega'') T_0^{hh}(x,E,\Omega,E,\Omega)\delta_{hh''}$$

$$+ \sum_{h'=1}^{H} c^{h'h''}(x,E'')/\mu'' \int_0^\infty dE' \int_0^{4\pi} d\Omega' f^{h'h''}(x,E',\Omega',E'',\Omega'')$$

$$\cdot T_M^{hh'}(x,E,\Omega,E',\Omega')] \qquad (h,h_0 = 1,2,\cdots,H). \tag{9.5-7}$$

The initial condition for Eq. (9.5-5) is unity while for the others it is zero. It is noted that Eq. (9.5-6) includes the unlikely series of interactions of passing through particle type-h as an intermediate and then having it reappear as the final result. Naturally, a degradation in energy would be expected with such a happening.

This apparent excessive amount of detail concerning various orders of scatterings is many times necessary in finite media problems where the largest contributions tend to be the unscattered components. The separation shown by these equations allows the numerical solution of the equations to be performed in an easier as well as lower order procedure.

9.6. AN EXAMPLE: THE CAPTURE GAMMA SHIELDING PROBLEM

As a brief example of the use of this multiparticle formulation, the capture gamma-ray problem for a nuclear reactor shield will be formulated. In this situation a neutron distribution is incident upon a shield; however, the atoms of the shield capture neutrons producing gamma-rays in the process. These gamma-rays can be of sufficient energy to be quite penetrating and thus escape the shield. A basic problem in shield design is to compensate for this capture gamma source of photons.

Let particle number one be the neutrons and particle number two be the photons; therefore, the quantity of interest becomes the 21-transmission function or for the known initial neutron source, $S(E_0, \Omega_0)$

$$T_{cg}(x,E) = \int_0^\infty dE_0 \int_0^{4\pi} d\Omega_0\, S(E_0,\Omega_0) \int_0^{4\pi} d\Omega\, T^{21}(x,E,\Omega,E_0,\Omega_0), \quad (9.6\text{-}1)$$

where the energy distribution of the transmitted photons remains but the angular distribution has been integrated out.

In formulating this problem the assumption is made that c^{12} is zero, that is no neutrons are produced as the result of gamma-ray interaction. Naturally, if a material like beryllium is present this assumption cannot be made.

Letting c^{12} be zero allows the like equations for the reflection and transmission to be omitted also. Then the equations necessary conceptually are R^{11}, R^{21}, R^{22}, T^{11}, T^{21}, T^{22}; however, the 11- and 22-reflection and transmission functions are for pure neutrons and gammas respectively. The neutron result has been shown in earlier chapters, while the gamma-ray form is discussed in Chapter 10. It is noted that the expected angular distribution of the neutrons and photons need not be represented by the same order quadrature formula in the discrete representation. In fact for the physically important problem the neutrons will be deep-penetrating ones such that a high forward component is necessary whereas the gamma photons will be near their creation point and being of near isotropic emittance a less peaked distribution is expected. The cross sections giving the resultant energy–angular distributions can be found on the general neutron cross-section tapes such as ENDF/B, see Honeck (1967), although the energy groups usually will need to be compiled into a broad group structure for these calculations.

To transform the equations into discrete form a nomenclature problem is present because of the large number of subscripts and superscripts necessary; therefore, the particle types will be shown as a superscript preceding the symbol. So that for slabs

$$R^{hh_0}(x,E_m,\mu_i,E_n,\mu_j) = {}^{hh_0}R^{ij}_{mn}(x). \quad (9.6\text{-}2)$$

The 21-equations are

$$\frac{d}{dx} {}^{21}R^{ij}_{mn}(x) = - \ [\Sigma^1_n(x)/\mu_j + \Sigma^2_m(x)/\mu_i] \ {}^{21}R^{ij}_{mn}(x)$$

$$+ \ \Sigma^1_n(x) \ c^{21}_n(x) \ {}^{21}f^{ij}_{mn}(x)/\mu_j$$

$$+ \ \Sigma^1_n(x)/\mu_j \ c^{11}_n(x) \sum_{p=1}^{N} \Delta E_p \sum_{s=1}^{L} w_s \ {}^{11}f^{sj}_{pn}(x) \ {}^{21}R^{is}_{mp}(x)$$

$$+ \ \Sigma^1_n(x)/\mu_j \ c^{21}_n(x) \sum_{p=1}^{N} \Delta E_p \sum_{s=1}^{L} w_s \ {}^{21}f^{sj}_{pn}(x) \ {}^{22}R^{is}_{mp}(x)$$

$$+ \sum_{q=1}^{N} \Delta E_q \ \Sigma^1_q(x) \ c^{21}_q(x) \sum_{u=1}^{L} w_u \ {}^{11}R^{uj}_{qn}(x) \ {}^{21}f^{iu}_{mq}(x)/\mu_u$$

$$+ \sum_{q=1}^{N} \Delta E_q \ \Sigma^2_q(x) \ c^{22}_q(x) \sum_{u=1}^{L} w_u \ {}^{21}R^{uj}_{qn}(x) \ {}^{22}f^{iu}_{mq}(x)/\mu_u$$

$$+ \sum_{q=1}^{N} \Delta E_q \ \Sigma^1_q(x) \sum_{u=1}^{L} w_u \ {}^{11}R^{uj}_{qn}(x)/\mu_u \ \{c^{11}_q(x) \sum_{p=1}^{N} \Delta E_p$$

$$\cdot \sum_{s=1}^{L} w_s \ {}^{11}f^{su}_{pq}(x) \ {}^{21}R^{is}_{mq}(x) + c^{21}_q(x) \sum_{p=1}^{N} \Delta E_p$$

$$\cdot \sum_{s=1}^{L} w_s \ {}^{21}f^{su}_{pq}(x) \ {}^{22}R^{is}_{mq}(x)\}$$

$$+ \sum_{q=1}^{N} \Delta E_q \ \Sigma^2_q(x) \sum_{u=1}^{L} w_u \ {}^{21}R^{uj}_{qn}(x)/\mu_u \ c^{22}_q(x) \sum_{p=1}^{N} \Delta E_p$$

$$\cdot \sum_{s=1}^{L} w_s \ {}^{22}f^{su}_{pq}(x) \ {}^{22}R^{is}_{mq}(x), \tag{9.6-3}$$

$$\frac{d}{dx} {}^{22}T^{ii}_{mmo}(x) = - \ \Sigma^2_m(x)/\mu_i \ {}^{22}T^{ii}_{mmo}(x), \tag{9.6-4}$$

and

$$\frac{d}{dx} {}^{21}T^{ij}_{mnS}(x) = - \ \Sigma^1_n(x)/\mu_j \ {}^{21}T^{ij}_{mnS}(x)$$

$$+ \ \Sigma^1_n(x) \ c^{21}_n(x)/\mu_j \ {}^{21}f^{ij}_{mn}(x) \ {}^{22}T^{ii}_{mmo}(x)$$

$$+ \ \Sigma^1_n(x) \ c^{11}_n(x)/\mu_j \sum_{p=1}^{N} \Delta E_p \sum_{s=1}^{L} w_s \ {}^{11}f^{sj}_{pn}(x) \ {}^{21}T^{is}_{mpS}(x)$$

$$+ \ \Sigma^1_n(x) \ c^{21}_n(x)/\mu_j \sum_{p=1}^{N} \Delta E_p \sum_{s=1}^{L} w_s \ {}^{21}f^{sj}_{pn}(x) \ {}^{22}T^{is}_{mpS}(x)$$

$$+ \sum_{q=1}^{N} \Delta E_q \sum_{u=1}^{L} w_u \, {}^{11}R_{qn}^{uj}(x) \, \Sigma_q^1(x) \, \{c_q^{21}(x)/\mu_u \, {}^{21}f_{mq}^{iu}(x) \, {}^{22}T_{mmo}^{ii}(x)$$

$$+ c_q^{11}(x)/\mu_u \sum_{p=1}^{N} \Delta E_p \sum_{s=1}^{L} w_s \, {}^{11}f_{pq}^{su}(x) \, {}^{21}T_{mpS}^{is}(x)$$

$$+ c_q^{21}(x)/\mu_u \sum_{p=1}^{N} \Delta E_p \sum_{s=1}^{L} w_s \, {}^{21}f_{pq}^{su}(x) \, {}^{22}T_{mpS}^{is}(x)\}$$

$$+ \sum_{q=1}^{N} \Delta E_q \sum_{u=1}^{L} w_u \, {}^{21}R_{qn}^{uj}(x) \, \Sigma_q^2(x) \, \{c_q^{22}(x)/\mu_u \, {}^{22}f_{mq}^{iu}(x) \, {}^{22}T_{mmo}^{ii}(x)$$

$$+ c_q^{22}(x)/\mu_u \, {}^{22}f_{mq}^{iu}(x) \, {}^{22}T_{mpS}^{is}(x)\}. \tag{9.6-5}$$

Here T_s refers to $T_1 + T_M$ and only the pure particle unscattered component has been broken out separately. The normal 11-equations and 22-equations are Eq. (5.2-10) and Eqs. (6.6-3,4).

These coupled equations although appearing to be quite complicated are computationally straightforward to solve in conjunction with the pure neutron and photon equations. It is noted that the provision for spatially dependent cross sections, such as for a layered shield, has been retained in the equations. This provision also allows an optimization-type problem to be formulated wherein the properties of the shield are varied so as to produce the minimum amount of transmitted gamma-rays. The size of this system of coupled ordinary first order differential equations may become large since the size not counting the unscattered transmission equations is

$$N_s = \{4[N(N+1)/2] + 2N^2\} L^2, \tag{9.6-6}$$

where down-scattering only has been utilized for the pure particle equations. For instance for $N = 10$ and $L = 4$, this is 6,720 equations which is a formidable task. Again the quadrature orders do not need to be the same so that in Eq. (9.6-3) and Eq. (9.6-5) the summation symbols u and s need not necessarily have the same total order; however, a careful rederivation of these equations would be necessary to determine the proper order for each summation. The same is true for the energy group structures. For details see the next section.

9.7. THE DISCRETE FORMS

To complete the picture the discrete forms for the case will be listed where N will refer to the total energy groups for particle type h, N_0 to h_0, and so forth. For the angular quadrature L will refer to h, L_0 to h_0, etc. Thus an energy summation limit of N' will refer to the corresponding value for the particle type h' since it appears inside of the particle summation. The discrete nomenclature as well as the scattering decomposition of Section 9.6 will be used.

The reflection equation then is for slab geometry

$$\frac{d}{dx}\,{}^{hh_0}R^{ij}_{mn}(x) = -\,[\Sigma^{h_0}_n(x)/\mu_j + \Sigma^h_m(x)/\mu_i]\,{}^{hh_0}R^{ij}_{mn}(x)$$

$$+\,\Sigma^{h_0}_n(x)\,c^{hh_0}_n(x)\,{}^{hh_0}f^{ij}_{mn}(x)/\mu_j$$

$$+\,\Sigma^{h_0}_n(x)/\mu_j\,\sum_{h'=1}^{H}\,c^{h'h_0}_n(x)\,\sum_{p=1}^{N'}\Delta E_p\,\sum_{s=1}^{L'}w_s\,{}^{h'h_0}f^{sj}_{pn}(x)\,{}^{hh'}R^{is}_{mp}(x)$$

$$+\,\sum_{h''=1}^{H}\,\sum_{q=1}^{N''}\Delta E_q\,\Sigma^{h''}_q(x)\,c^{hh''}_q(x)\,\sum_{u=1}^{L''}w_u\,{}^{h''h_0}R^{uj}_{qn}(x)\,{}^{hh''}f^{iu}_{mq}(x)/\mu_u$$

$$+\,\sum_{h''=1}^{H}\,\sum_{q=1}^{N''}\Delta E_q\,\Sigma^{h''}_q(x)\,\sum_{u=1}^{L''}w_u\,{}^{h''h_0}R^{uj}_{qn}(x)/\mu_u\,\sum_{h'=1}^{H}c^{h'h''}_q(x)$$

$$\cdot\,\sum_{p=1}^{N'}\Delta E_p\,\sum_{s=1}^{L'}w_s\,{}^{h'h''}f^{su}_{pq}(x)\,{}^{hh'}R^{is}_{mp}(x), \tag{9.7-1}$$

while the transmission equations become

$$\frac{d}{dx}\,{}^{h_0h_0}T^{ii}_{mmo}(x) = -\,\Sigma^{h_0}_n(x)/\mu_j\,{}^{h_0h_0}T^{ii}_{mmo}(x), \tag{9.7-2}$$

$$\frac{d}{dx}\,{}^{hh_0}T^{ij}_{mnS}(x) = -\,\Sigma^{h_0}_n(x)/\mu_j\,{}^{hh_0}T^{ij}_{mnS}(x)$$

$$+\,\Sigma^{h_0}_n(x)\,c^{hh_0}_n(x)/\mu_j\,{}^{hh_0}f^{ij}_{mn}(x)\,{}^{hh}T^{ii}_{mmo}(x)$$

$$+\,\Sigma^{h_0}_n(x)\,\sum_{h'=1}^{H}c^{h'h_0}_n(x)/\mu_j\,\sum_{p=1}^{N'}\Delta E_p\,\sum_{s=1}^{L'}w_s\,{}^{h'h_0}f^{sj}_{pn}(x)\,{}^{hh'}T^{is}_{mpS}(x)$$

$$+\,\sum_{h''=1}^{H}\,\sum_{q=1}^{N''}\Delta E_q\,\sum_{u=1}^{L''}w_u\,{}^{h''h_0}R^{uj}_{qn}(x)\,\Sigma^{h''}_q(x)\,\{c^{hh''}_q(x)/\mu_u$$

$$\cdot\,{}^{hh''}f^{iu}_{mq}(x)\,{}^{hh}T^{ii}_{mmo}(x) + \sum_{h'=1}^{H}c^{h'h''}_q(x)/\mu_u\,\sum_{p=1}^{N'}\Delta E_p$$

$$\cdot\,\sum_{s=1}^{L'}w_s\,{}^{h'h''}f^{su}_{pq}(x)\,{}^{hh'}T^{is}_{mpS}(x)\} \qquad (h,h_0 = 1,2,\cdots,H;i = 1,2,\cdots,L;$$

$$j = 1,2,\cdots,L_0;m = 1,2,\cdots,N;n = 1,2,\cdots,N_0). \tag{9.7-3}$$

The initial condition for all equations but Eq. (9.7-2) is zero. Equation (9.7-2) has the initial condition of unity.

9.8. INTERNAL SOURCE PROBLEMS

The equations for internal source problems involving the E- and G-functions are not considered here since in coupled particle problems, one type of particle acts as a source for other types so that only reflection and transmission results are necessary. However, conceptually the E- and G-functions could be formulated for multiparticle use if desired.

9.9. EXERCISES

1. Deduce a physical example whereby in Eq. (9.3-1) the nonlinear term could represent a hh_0-reflection followed by an $h_0 h$-interaction followed by an hh_0-reflection.

2. Draw the interaction diagram similar to Figure 4 for Eqs. (9.5-1,2,3).

3. Draw the interaction diagram similar to Figure 6 for Eqs. (9.5-5,6,7).

4. Expand Eq. (9.6-3) into the $R_1 + R_M$ form.

5. Expand Eq. (9.6-5) into the $T_1 + T_M$ form.

6. Explain the form of Eq. (9.6-6).

7. Set up the problem to determine the reflection of neutrons from a natural beryllium reflector where both incident neutrons as well as gamma rays are present. Include the following reactions as special cases and count the reflected neutrons from each reaction as a separate particle.

$$Be^9 (n, \, 2n) \, 2He^4$$

$$Be^9 (\gamma, n) \, 2He^4$$

In addition the standard scattering and normal capture reactions can occur.

CHAPTER 10

THE SPECIAL CASE OF PHOTON TRANSPORT

10.1 INTRODUCTION

The transport of photons, in particular high energy photons, within a medium involves some special considerations because of the particular form of the physical scattering laws. The properties of the "photoelectric effect," "Compton scattering," and "pair production" are explained in the next section where the appropriate cross sections are determined. In addition, transport theory for photons is much more concerned with the energy distribution of the photons rather than the spatial distribution. In contrast neutron transport is usually concerned greatly with the spatial neutron distribution particularly as it affects the leakage from a medium. Consequently, for photons the semiinfinite slab problem involving the distribution of energy, or its equivalent in wavelength, of the reflection function is an important result.

Another change in photon transport theory is that the usual nomenclature is different from that utilized previously. The nomenclature of this book has a tendency to follow the field of neutron transport theory so that in this chapter footnotes will be employed to explain the equivalence in nomenclature.

10.2. THE CROSS SECTIONS FOR PHOTONS

In nuclear reactor technology the energy range of interest for photons is from a few hundred kiloelectron volts (kev) to ten to twenty million electron volts (Mev). In this range the interaction of photons with the electrons of nuclei are by the photoelectric effect, Compton scattering, and pair production. The photoelectric effect gets its name from the fact that a photon is absorbed near an atomic nucleus so that its energy is transferred to an orbital electron, thus producing a free electron. Certain chemical compounds such as cadmium sulfide show this effect with photons of the energy range of the visual wavelengths and the resulting electrical current is referred to as photoelectric current. Also resulting from a photoelectric absorption are soft x-rays caused by the orbital electrons filling in the "space" left by the leaving electron; however these x-rays are very low in energy compared to the other gamma-ray photons present and are normally ignored in the analysis as the photoelectric cross section for absorption becomes larger as the photon energy decreases. Since in the transport equation the absorption cross sections do not appear directly, it is common to obtain this photoelectric cross section by subtracting the other effects from the total cross section.

In the intermediate energy ranges under consideration the Compton scattering effect takes place. This is an elastic collision between a photon and an electron resulting in a free electron being created and a degraded photon traveling in a new direction. Mathematically this energy conservation is described by

$$\lambda = \lambda_0 + 1 - \cos\theta_s, \tag{10.2-1}$$

where λ is the energy expressed in Thompson units or

$$\lambda = mc^2/E. \tag{10.2-2}$$

Here mc^2 is the rest-*mass* energy of an electron, or 0.51 Mev, E is the energy of the photon, and θ_s is the angle of deflection of the photon. The subscript zero is for the initial photon and no subscript is the scattered photon. The scattering angle can be expressed in terms of the initial and final angles in a spherical coordinate system as

$$\cos\theta_s = \mu\mu_0 + \sqrt{1-\mu_0^2}\sqrt{1-\mu^2}\cos(\phi - \phi_0), \tag{10.2-3}$$

where $\mu = \cos\theta$.* The probability for an interaction is given by the Klein-Nishisa cross section, see Fano, Spencer, and Berger (1959), as

$$\frac{d^2\sigma_s}{d\lambda d\Omega} = \frac{r_0^2}{2}K(\lambda_0,\lambda)\,\delta(1+\lambda_0-\lambda-\cos\theta_s), \tag{10.2-4}$$

where

$$K(\lambda_0,\lambda) = [\lambda_0/\lambda]^2\,[\lambda/\lambda_0 + \lambda_0/\lambda + (\lambda_0-\lambda)^2 + 2(\lambda_0-\lambda)] \tag{10.2-5}$$

and $r_0 = 2.82\,(10^{-13})$ cm and is the classical atomic radius. This cross section is naturally for a single electron present in the medium and is multiplied by the appropriate electron density, N_e, when actually employed in the transport equation. Thus

$$\Sigma_s = N_e\sigma_s. \tag{10.2-6}$$

In slab geometry the integration over the ϕ-dependence as is shown in Chapter 4 can be expressed as

$$\frac{d^2\sigma_s}{d\lambda d\mu} = \frac{1}{2}r_0^2\,K(\lambda_0,\lambda)\int_{\mu\mu_0+\sqrt{1-\mu_0^2}\sqrt{1-\mu^2}}^{\mu\mu_0-\sqrt{1-\mu_0^2}\sqrt{1-\mu^2}} d(\cos\theta_s)$$

$$\cdot\frac{\delta(1+\lambda_0-\lambda-\cos\theta_s)}{[1-\mu_0^2-\mu^2-\cos^2\theta_s+2\mu\mu_0\cos\theta_s]^{1/2}} \tag{10.2-7}$$

*The common symbol for μ is ω in photon transport literature.

provided that neither μ nor μ_0 is unity. It is to be noted that between the limits of integration that the δ-function values of $\cos\theta_s$ appear twice, and thus two equal contributions to the integration will be generated. The result is

$$\frac{d^2\sigma_s}{d\lambda d\mu} = r_0^2 K(\lambda_0,\lambda) [1 - \mu_0^2 - \mu^2 - \gamma^2 + 2\mu\mu_0\gamma]^{-1/2}, \qquad (10.2\text{-}8)$$

where

$$\gamma = 1 + \lambda_0 - \lambda. \qquad (10.2\text{-}9)$$

The total scattering microscopic cross section then is

$$\sigma_s(\lambda_0) = \int_{\lambda_0}^{\lambda_0+2} d\lambda \int_{-1}^{+1} d\mu \frac{d^2\sigma_s}{d\lambda d\mu} = r_0^2\lambda_0 \left\{ (1 - 2\lambda_0 - 2\lambda_0^2) \ln(1 + 2/\lambda_0) \right.$$

$$\left. + 2(1 + 9\lambda_0 + 8\lambda_0^2 + 2\lambda_0^3)(\lambda_0 + 2)^{-2} \right\}. \qquad (10.2\text{-}10)$$

In terms of the quantities used in the invariant imbedding transport equations of previous chapters the scattering matrix is from Eq. (4.3-1)

$$[\Sigma_n c_n f_{mn}(\mu_i,\mu_j)]_{KN} = \frac{N_e}{\Delta E_n} \int_{E_n}^{E_{n+1}} dE_0 \int_{E_m}^{E_{m+1}} dE \frac{d^2\sigma_s}{d\lambda d\mu}$$

$$= N_e r_0^2 \frac{\lambda_n \lambda_{n+1}}{\lambda_{n+1} - \lambda_n} \int_{\lambda_n}^{\lambda_{n+1}} d\lambda_0$$

$$\cdot \int_{\lambda_m}^{\lambda_{m+1}} d\lambda \frac{K(\lambda_0,\lambda)}{\lambda_0^2 [1 - \mu_0^2 - \mu^2 - \gamma^2 + 2\mu\mu_0\gamma]^{1/2}}. \qquad (10.2\text{-}11)$$

The λ-integration can be performed analytically while the λ_0-integration is done numerically.

The analytical portion of the integration is performed by rearranging Eq. (10.2-11) into forms like

$$I_s = \frac{g_s(\lambda_0)}{\lambda^s [A\lambda^2 + B\lambda + C]^{1/2}}, \qquad (10.2\text{-}12)$$

where

$$g_0(\lambda_0) = \lambda_0^2, \qquad (10.2\text{-}13)$$

$$g_1(\lambda_0) = \lambda_0(1 - 2\lambda_0 - 2\lambda_0^2), \qquad (10.2\text{-}14)$$

$$g_2(\lambda_0) = \lambda_0^3(\lambda_0 + 2), \qquad (10.2\text{-}15)$$

$$g_3(\lambda_0) = \lambda_0^3, \qquad (10.2\text{-}16)$$

$$A = -1, \qquad (10.2\text{-}17)$$

$$B = 2(1 + \lambda_0 - \mu\mu_0), \qquad (10.2\text{-}18)$$

THE ELMER E. RASMUSON LIBRARY
UNIVERSITY OF ALASKA

and

$$C = -\mu_0^2 - \mu^2 - \lambda_0(2 + \lambda_0) + 2\mu\mu_0(1 + \lambda_0). \qquad (10.2\text{-}19)$$

The integral for $s = 0$ is an arcsin while those for $s > 0$ depend upon the sign of C so that

$$I_0 = g_0(\lambda_0) \sin^{-1} \frac{2\lambda + B}{\sqrt{B^2 + 4C}}, \qquad (10.2\text{-}20)$$

$$I_1 = g_1(\lambda_0) \cdot \begin{cases} -\dfrac{1}{\sqrt{C}} \ln\left\{ \dfrac{\sqrt{-\lambda^2 + B\lambda + C} + \sqrt{C}}{\lambda} + \dfrac{B}{2\sqrt{C}} \right\} & (C > 0) \\[3ex] -\dfrac{2}{B\lambda} \sqrt{-\lambda^2 + B\lambda} & (C = 0) \\[3ex] \dfrac{1}{\sqrt{-C}} \sin^{-1} \dfrac{B\lambda + 2C}{\lambda\sqrt{B^2 + 4C}} & (C < 0), \qquad (10.2\text{-}21) \end{cases}$$

and

$$I_s = \left\{ -\frac{\sqrt{-\lambda^2 + B\lambda + C}}{C(s-1)\lambda^{s-1}} + \frac{B(3-2s)}{2C(s-1)} I_{s-1} - \frac{(2-s)}{C(s-1)} I_{s-2} \right\} g_s(\lambda_0)$$

$$(s = 2,3), \qquad (10.2\text{-}22)$$

provided that the λ_m and λ_{m+1} lie between the roots of

$$-\lambda^2 + B\lambda + C = 0 \qquad (10.2\text{-}23)$$

thus assuring positive square roots everywhere. If this is not the case then the applicable portion of the range is allowed up to the corresponding root. If the $\lambda_{m+1} \leqslant \lambda \leqslant \lambda_m$ range lies entirely outside of the root values, then no physical transfer of this nature can be allowed and the integration is zero. In the special case of either μ or μ_0 being unity, the roots of Eq. (10.2-13) become equal and it is necessary to utilize Eq. (10.2-3) directly noting that now no ϕ-dependence appears. Thus the ϕ-integration can be done directly giving

$$\frac{d^2\sigma_s}{d\lambda d\mu} = \int_0^{2\pi} d\phi \frac{d^2\sigma_s}{d\lambda d\Omega} = \pi r_0^2 K(\lambda_0, \lambda^*), \qquad (10.2\text{-}24)$$

where

$$\lambda^* = 1 + \lambda_0 - \mu\mu_0 \qquad (\mu \text{ or } \mu_0 = 1). \qquad (10.2\text{-}25)$$

The numerical integration over λ_0 can be performed by any convenient and relatively accurate method. Simpson's rule is commonly employed because of its ease of programming. Each group range λ_n to λ_{n+1} is evaluated at an odd number, N_I, usually seven or larger, of values of λ_0 equally spaced; thus

$$[\Sigma_n c_n f_{mn}(\mu_i,\mu_j)]_{KN} = \frac{N_e r_0^2 \lambda_n \lambda_{n+1}}{\lambda_{n+1} - \lambda_n} \sum_{p=1}^{N_I} w_p \Delta \lambda_0$$

$$\cdot \sum_{s=0}^{4} [I_s(\lambda^{++},\lambda_{op}) - I_s(\lambda^+,\lambda_{op})], \qquad (10.2\text{-}26)$$

where

$$\lambda^{++} = \min(\lambda_2,\lambda_{m+1}), \qquad (10.2\text{-}27)$$

$$\lambda^+ = \max(\lambda_1,\lambda_m), \qquad (10.2\text{-}28)$$

$$w_p = \begin{cases} 1 & (p = 1, N_I) \\ 4 & (\mathrm{mod}(p,2) = 0) \\ 2 & (\mathrm{mod}(p + 1,2) = 0), \end{cases} \qquad (10.2\text{-}29)$$

and

$$\Delta \lambda_0 = \frac{\lambda_{n+1} - \lambda_n}{N_I - 1}. \qquad (10.2\text{-}30)$$

Here λ_2 is the largest root of Eq. (10.2-23) and λ_1 is the smallest. If μ or μ_0 is unity, then Eq. (10.2-24) is utilized in place of the last s-summation.

If a conservative cross section matrix is desired then the integration interval of the lowest energy group must extend so that

$$\lambda_N - \lambda_{N-1} = 2 \qquad (10.2\text{-}31)$$

which is the maximum energy change for one scattering and corresponds to a purely backscattered photon.

For additional details about photon scattering theory see Kaiser (1968).

In the high energy range above two electron rest-masses, or 1.02 Mev, the phenomenon of "pair production" occurs. In this interaction process the photon under the influence of the forces surrounding an atom is transformed into an electron and a positron moving in near opposite directions. Each of these particles has the rest-mass of an electron but the electron is negatively charged while the positron is positively charged. Any additional excess energy that the photon had over and above the creation energy shows up in the kinetic energy of these two particles. However, since the particles are charged their range before slowing down to negligible speed is very limited, especially in comparison to the large mean free path of the photons. During this slowing down process the emission of Bremsstrahlung radiation, i.e., continuous low energy x-rays, is observed because the charged particle is moving in a curved path in an electric field. This Bremsstrahlung is low enough in energy to be normally ignored as far as radiation shielding is concerned. On the other hand, once the

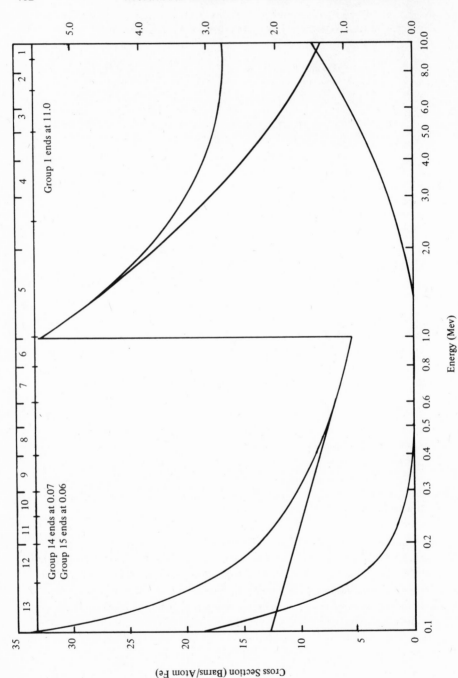

Fig. 8. Photon cross sections for iron.

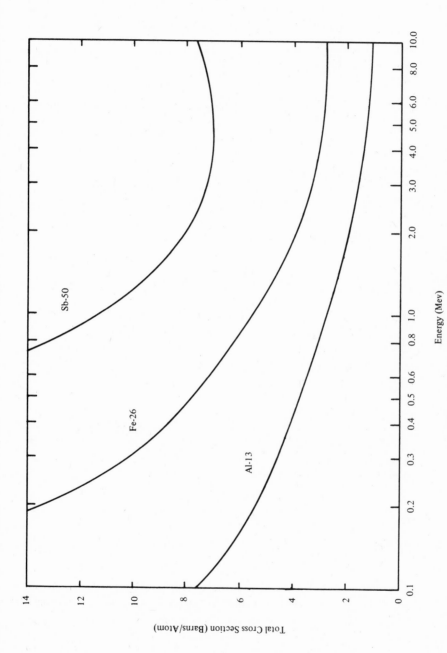

Fig. 9. Total photon cross sections for selected materials.

positron has lost sufficient kinetic energy it is "ripe" for an interaction with another electron. This annihilation interaction where matter and antimatter combine results in "pure energy" being created in the form of two annihilation photons moving in opposite directions and each having an energy of one electron rest-mass, or 0.51 Mev. It is usually assumed that these annihilation photons are created at the same position as the original pair-production interaction and thus any positron travel is ignored. Therefore the result of a pair production interaction is a degradation of the photon energy from one high energy to two 0.51 Mev photons moving in opposite directions to each other but randomly oriented in comparison to the original photon direction.

In terms of cross sections the pair production phenomenon can be modeled as

$$f(E,\mu,E_0,\mu_0)dEd\mu = \delta(E - 0.51 \text{ Mev})dE \, d\mu/2 \qquad (10.2\text{-}32)$$

and the c-value is two. In discrete form this is

$$\Sigma_n c_n f^{ij}_{mn} = \frac{1}{2} \Sigma^p_n \delta_{mm_p} \{ \delta_{ik_p} + \delta_{ik'_p} \} \qquad (10.2\text{-}33)$$

where Σ^p_n is the macroscopic cross section* for pair production for photons in the energy range λ_n to λ_{n+1}, m_p is the dimensionless energy group containing 0.51 Mev and k_p and k'_p are oppositely oriented angle indices.

In many instances it is convenient to follow the annihilation photons separately and thus treat them as a separate particle type. Therefore, if a type-1 particle is the original photon and type-2 particle the annihilation photon, then R^{21} is the annihilation photon albedo and R^{11} and R^{22} are both albedoes for photons except that R^{22} has an upper energy range of 0.51 Mev.

The three contributions to the total cross section for photons are shown in Fig. 8 for iron while Fig. 9 shows the total cross section for several materials obtained from the data of Storm (1967). It is noted that the cross sections are larger for heavy atoms, i.e., large Z-numbers.

10.3. THE ALBEDO FOR PHOTONS

Using the development for the scattering and absorption laws as explained in the previous section, the albedo for nonannihilation photons is given by Eq. (5.3-15) and Eq. (5.3-16) where Σ_n is the average total cross section for the energy range λ_n to λ_{n+1} and Eq. (10.2-11) gives the scattering contributions. For the annihilation albedo the R^{21} results, as shown by Eq. (9.7-1) when transformed to the $\mathcal{R}$-form by Eq. (5.3-13,14), can be used where R^{11} and R^{22} are

---

*The common symbol for Σ^p is $N\kappa$ in photon transport literature.

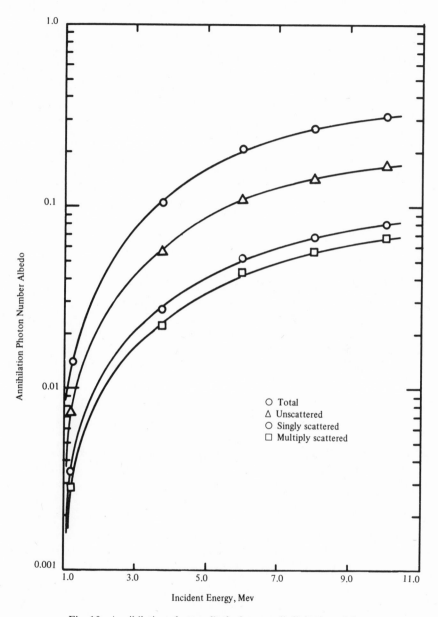

Fig. 10. Annihilation photon albedo for a semiinfinite iron slab.

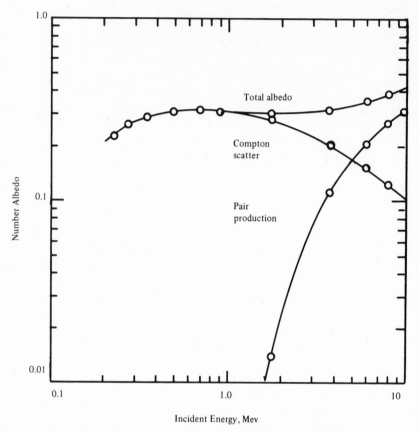

Fig. 11. Incident energy dependence of the total albedo for a semiinfinite iron slab.

identical and c^{12} is assumed zero. Usually the expanded form involving $R_1 + R_m$ is utilized in the calculations especially if a homogeneous slab is utilized so that R_1 has an analytical solution.

Figure 10 gives the annihilation photon albedo, R^{21}, for a semiinfinite iron slab and shows the unscattered, single scattered, and multiscattered components. Figure 11 shows the total albedo for the same semiinfinite iron slab showing the scattering and pair production contributions. These calculations are performed with a fifteen group energy set, which is shown in Fig. 8, and a Gaussian quadrature of order four.

10.4 OTHER PHOTON RESULTS

The calculations for the transmission of photons through various materials have been performed by Kaiser (1968) and Shimizu (1967); however, the latter results give only the total dose effects and thus the energy-angle dependence is not shown.

10.5 EXERCISES

1. Show the use of the semiinfinite albedo results when solving for the finite slab albedo by noting that the mean free path for the low and high energy photons differs by an order of magnitude. What other numerical integration problems are imposed by this large mean-free-path variance?

2. In the energy range of photons of 10 Mev or higher, the positron, when it is created by pair-production, is energetic enough that it travels a significant distance before it is annihilated. If this "positron travel" is given by a range factor, $R(E)$ where E is the energy of the photon when pair-production occurs, derive the applicable reflection equations if pair-production is assumed to be isotropic in nature.

CHAPTER 11

CURVED GEOMETRY CONSIDERATIONS

11.1 INTRODUCTION

In the previous chapters it has been assumed that the geometry of the transport material has been a slab. In this chapter the ramifications necessary for curved geometries are discussed. In particular the case of spherical geometry is analyzed and the difficulties outlined.

11.2 GENERAL CURVED GEOMETRY APPROACHES

In reference to the invariant imbedding derivation in Section 3.3 for the reflection function, the following statement is found:

$$R(x + dx, E, \Omega, E_0, \Omega_0)\, dE\, d\Omega = \text{sum of five terms.}$$

These five terms on the right side are of order dx except for one term that cancels with the first term of the Taylor's series expansion of the above. In curved geometries the same situation occurs except that the Taylor's series expansion must include all effects, and particularly the change in solid angle with a straight path through the medium is important. In addition the change in the differential solid angle must be included. This latter term has given considerable difficulty in deriving these curved geometry equations as has been indicated by Bailey and Wing (1964).

In general the expansion can be represented as

$$R(r + dr, E, \Omega, E_0, \Omega_0)dE\, d\Omega = R(r, E, \Omega, E_0, \Omega_0)\, d(E + dE)d(\Omega + d\Omega)$$

$$+ \frac{\partial}{\partial r} R(r, E, \Omega, E_0, \Omega_0)dE\, d\Omega + \frac{\partial}{\partial E} R(r, E, \Omega, E_0, \Omega_0)\frac{dE}{dr} dE\, d\Omega$$

$$+ \frac{\partial}{\partial \Omega} R(r, E, \Omega, E_0, \Omega_0)\frac{d\Omega}{dr} dE\, d\Omega + \frac{\partial}{\partial E_0} R(r, E, \Omega, E_0, \Omega_0)\frac{dE_0}{dr} dE\, d\Omega$$

$$+ \frac{\partial}{\partial \Omega_0} R(r, E, \Omega, E_0, \Omega_0)\frac{d\Omega_0}{dr} dE\, d\Omega. \tag{11.2-1}$$

In neutron and gamma-ray transport theory the dE/dr terms are zero since the particle does not change energy while moving in a straight path with no interactions. However, for charged particles such as electrons the dE/dr term would be nonzero as an energy loss can be obtained without a definite interaction. The $d\Omega/dr$ term is nonzero in a curved geometry as a change in angle occurs while the particle is moving in a straight path. The $d(\Omega + d\Omega)$ term while appearing to be of higher differential order in reality has a term of order dr associated with it and thus contributes to the equation. In the next section this term is derived for spherical geometry.

11.3. THE CASE OF SPHERICAL GEOMETRY

The case of a uniform spherical region with angular symmetry about the center allows only radial and the usual energy and angular dependence. Thus, define

$R(r, E, \Omega, E_0, \Omega_0)\, dE\, d\Omega$ = the reflected particle current at r in a spherical region in the energy range dE about E and the direction $d\Omega$ about Ω because of a unit current input at the energy E_0 in the direction Ω_0.

Note that in a spherical region the reflection function is all that is necessary if the sphere is solid; however, in a spherical shell the center region determines whether a type of transmission function can be defined. For instance, a central region having total absorption allows a transmission definition. On the other hand, a vacuum central region allows streaming and thus the particles reenter the medium. This boundary would have the same angular energy distribution leaving as entering on the inside of this spherical region.

In this introduction to curved geometries the simplest case of utilizing a solid sphere will be used in order to show the geometry effects. Thus only the reflection function is needed for a symmetric sphere.

In reference to Section 3.2 and Eq. (11.2-1) the reflection function becomes

$$\frac{\partial}{\partial r} R(r, E, \Omega, E_0, \Omega_0) + \frac{\partial}{\partial \Omega} R(r, E, \Omega, E_0, \Omega_0)\frac{d\Omega}{dr}$$

$$+ \frac{\partial}{\partial \Omega_0} R(r, E, \Omega, E_0, \Omega_0)\frac{d\Omega_0}{dr} + R(r, E, \Omega, E_0, \Omega_0)\frac{\partial}{\partial \Omega}\left(\frac{d\Omega}{dr}\right)$$

$$+ \left[\Sigma(r, E_0)/|\mu_0| + \Sigma(r, E)/\mu\right] R(r, E, \Omega, E_0, \Omega_0)$$

$$= \Sigma(r, E_0)c(r, E_0)f(r, E, \Omega, E_0, \Omega_0)/|\mu_0|$$

$$+ \Sigma(r, E_0)c(r, E)/|\mu_0| \int_0^\infty dE' \int_0^{4\pi} d\Omega'\, f(r, E', \Omega', E_0, \Omega_0)$$

$$\cdot R(r, E, \Omega, E', \Omega')$$

$$+ \int_0^\infty dE'' \, \Sigma(r, E'') c(r, E'') \int_0^{4\pi} d\Omega'' \, R(r, E'', \Omega'', E_0, \Omega_0)$$
$$\cdot f(r, E, \Omega, E'', \Omega'')/\mu''$$

$$+ \int_0^\infty dE'' \, \Sigma(r, E'') c(r, E'') \int_0^\infty dE' \int_0^{4\pi} d\Omega'' \int_0^{4\pi} d\Omega' \, R(r, E'', \Omega'', E_0, \Omega_0)$$
$$\cdot f(r, E', \Omega', E'', \Omega'') R(r, E, \Omega, E', \Omega')/\mu'' \qquad (11.3\text{-}1)$$

Since a symmetric input with respect to the ϕ-dependence is assumed, this variable is integrated out by

$$R(r, E, \mu, E_0, \mu_0) dE \, d\mu = \frac{1}{2\pi} \int_0^{2\pi} d\phi_0 \int_0^{2\pi} d\phi \, R(r, E, \Omega, E_0, \Omega_0) dE \, d\mu$$

$$(11.3\text{-}2)$$

In addition, to show the geometry effects as simply as possible, the case of monoenergetic particles is employed by using

$$R(r, \mu, \mu_0) d\mu = \int_0^\infty dE \, R(r, E, \mu, E_0, \mu_0) d\mu \, \delta(E - E_0). \qquad (11.3\text{-}3)$$

Thus Eq. (11.3-1) reduces to

$$\frac{\partial}{\partial r} R(r, \mu, \mu_0) + \frac{\partial}{\partial \mu} R(r, \mu, \mu_0) \frac{d\mu}{dr}$$

$$+ \frac{\partial}{\partial \mu_0} R(r, \mu, \mu_0) + R(r, \mu, \mu_0) \frac{\partial}{\partial \mu}\left(\frac{d\mu}{dr}\right)$$

$$+ [\Sigma(r)/|\mu_0| + \Sigma(r)/\mu] \, R(r, \mu, \mu_0) = \Sigma(r) c(r) f(r, \mu, \mu_0)/|\mu_0|$$

$$+ \Sigma(r) c(r)/|\mu_0| \int_{-1}^0 d\mu' \, f(r, \mu', \mu_0) R(r, \mu, \mu')$$

$$+ \Sigma(r) c(r) \int_0^1 d\mu'' \, R(r, \mu'', \mu_0) f(r, \mu, \mu'')/\mu''$$

$$+ \Sigma(r) c(r) \int_0^1 d\mu'' \int_{-1}^0 d\mu' \, R(r, \mu'', \mu_0) f(r, \mu', \mu'') R(r, \mu, \mu')/\mu''.$$

$$(11.3\text{-}4)$$

The quantity $d\mu/dr$ can be determined from Fig. 12 which shows the change in angle with position in a spherical region. This is

$$\frac{d\mu}{dr} = \frac{1 - \mu^2}{\mu r}. \qquad (11.3\text{-}5)$$

If the usual notation change is made so that the input angle is defined with respect to its normal inward direction, then μ_0 is positive and an equation similar to Eq. (11.3-5) is obtained. The extra quantity becomes

$$\frac{\partial}{\partial \mu}\left(\frac{1 - \mu^2}{\mu r}\right) = -\frac{1 + \mu^2}{\mu^2 r}. \qquad (11.3\text{-}6)$$

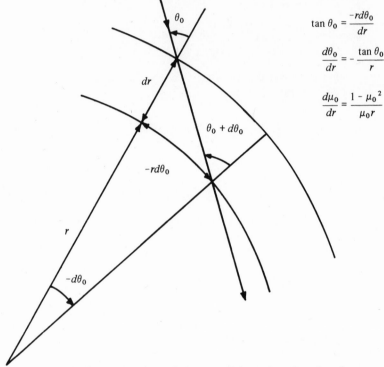

$$\tan\theta_0 = \frac{-rd\theta_0}{dr}$$

$$\frac{d\theta_0}{dr} = -\frac{\tan\theta_0}{r}$$

$$\frac{d\mu_0}{dr} = \frac{1-\mu_0{}^2}{\mu_0 r}$$

Fig. 12. Change in polar angle because of change in radius of a sphere.

Thus Eq. (11.3-4) reduces to

$$\left[\frac{\partial}{\partial r} + \frac{1-\mu^2}{\mu r}\frac{\partial}{\partial\mu} + \frac{1-\mu_0^2}{\mu_0 r}\frac{\partial}{\partial\mu_0} - \frac{1+\mu^2}{\mu^2 r} + \frac{1}{\mu} + \frac{1}{\mu_0}\right] R(r,\mu,\mu_0)$$

$$= cf(\mu,\mu_0)/\mu_0 + c/\mu_0 \int_0^1 d\mu' \, f(\mu',\mu_0)\, R(r,\mu,\mu')$$

$$+ c \int_0^1 d\mu'' \, R(r,\mu'',\mu_0)\, f(\mu,\mu'')/\mu''$$

$$+ c \int_0^1 d\mu'' \int_0^1 d\mu' \, R(r,\mu'',\mu_0)\, f(\mu',\mu'')\, R(r,\mu,\mu')/\mu'', \qquad (11.3\text{-}7)$$

where r is now mean free paths, i.e.,

$$\tilde{r} = \int_0^r \Sigma(r)dr \qquad (11.3\text{-}8)$$

and the ~-sign is then dropped for convenience. Also the spatial dependence has been deleted from the other properties.

Equation (11.3-7) represents the total reflection from a sphere with unit current input on the surface; therefore

$$\lim_{r \to 0} 4\pi r^2 \, R(r, \mu, \mu_0) = 1 \qquad (11.3\text{-}9)$$

is the initial condition for a solid sphere. In the case of a spherical shell with a fully absorbing center region, then

$$R(r_0, \mu, \mu_0) = 0, \qquad (11.3\text{-}10)$$

where r_0 is the radius of the inside of the shell. This condition is really only approximate since an uncollided contribution would still pass through the absorbing central sphere; however for practical r_0-values this contribution would be negligible. In order to exactly utilize Eq. (11.3-10) the central region can assume to be fully absorbing with a zero mean free path and thus all particles will be absorbed within a differential distance into the central medium.

Equation (11.3-7) can be changed to a conservative form of the equation by a rearrangement and collection of terms if the usual transformation

$$\hat{R}(r, \mu, \mu_0) = \mu_0 \, R(r, \mu, \mu_0) \qquad (11.3\text{-}11)$$

is made. This becomes

$$\frac{1}{r^2} \frac{\partial}{\partial r} [r^2 \, \hat{R}(r, \mu, \mu_0)] + \frac{1}{r} \frac{\partial}{\partial \mu} \left[\frac{1 - \mu^2}{\mu} \hat{R}(r, \mu, \mu_0) \right]$$

$$+ \frac{1}{r} \frac{\partial}{\partial \mu_0} \left[\frac{1 - \mu_0^2}{\mu_0} \hat{R}(r, \mu, \mu_0) \right] + \left[\frac{1}{\mu} + \frac{1}{\mu_0} \right] \hat{R}(r, \mu, \mu_0)$$

$$= c \, \{ f(\mu, \mu_0) + \int_0^1 d\mu' \, f(\mu', \mu_0) \hat{R}(r, \mu, \mu') / \mu'$$

$$+ \int_0^1 d\mu'' \, \hat{R}(r, \mu'', \mu_0) \, f(\mu, \mu'') / \mu''$$

$$+ \int_0^1 d\mu'' \, \hat{R}(r, \mu'', \mu_0) / \mu'' \int_0^1 d\mu' \, f(\mu', \mu'') \, \hat{R}(r, \mu, \mu') / \mu' \}.$$

$$(11.3\text{-}12)$$

It is noted that in this conservative form the previous extra term disappears as would be expected.

One of the problems associated with these invariant imbedding equations for a sphere is the contribution of the unscattered component. By geometrical considerations this is

$$R_0(r, \mu, \mu) = \exp(-2r\mu) \qquad (11.3\text{-}13)$$

but if the reflected current is broken into the unscattered and scattered components,

$$\hat{R}(r, \mu, \mu_0) = \hat{R}_0(r, \mu, \mu_0) \, \delta(\mu - \mu_0) + \hat{R}_s(r, \mu, \mu_0). \qquad (11.3\text{-}14)$$

The substitution of Eq. (11.3-14) into Eq. (11.3-12) does not appear to produce a differential equation whose solution is Eq. (11.3-13). A separate derivation based upon starting out with the unscattered and scattered components is shown by Mingle (1969) and produces for the additional restriction of isotropic scattering, i.e.,

$$f(\mu, \mu_0) \, d\mu = d\mu/2. \tag{11.3-15}$$

The equation

$$\frac{1}{r^2} \frac{\partial}{\partial r} \left[r^2 \, \hat{R}_s(r, \mu, \mu_0) \right] + \frac{1}{r} \frac{\partial}{\partial \mu} \left[\frac{1 - \mu^2}{\mu} \hat{R}_s(r, \mu, \mu_0) \right]$$

$$+ \frac{1}{r} \frac{\partial}{\partial \mu_0} \left[\frac{1 - \mu_0^2}{\mu_0} \hat{R}_s(r, \mu, \mu_0) \right] + \left[\frac{1}{\mu} + \frac{1}{\mu_0} \right] \hat{R}_s(r, \mu, \mu_0)$$

$$= \frac{c}{2} \left\{ 1 + \hat{R}_0(r, \mu, \mu)/\mu + \int_0^1 d\mu' \, \hat{R}_s(r, \mu, \mu')/\mu' \right\}$$

$$\cdot \left\{ 1 + \hat{R}_0(r, \mu_0, \mu_0)/\mu_0 + \int_0^1 d\mu'' \, \hat{R}_s(r, \mu'', \mu_0)/\mu'' \right\}. \tag{11.3-16}$$

Here

$$\hat{R}_0(r, \mu, \mu) = \mu \, R_0(r, \mu, \mu) \tag{11.3-17}$$

as would be expected. The boundary condition then is

$$\lim_{r \to 0} 4\pi r^2 \, \hat{R}_s(r, \mu, \mu_0) = 0 \tag{11.3-18}$$

for this scattered component.

The reflection equation for a spherical shell with a symmetric input has been analyzed by Bellman, Kagiwada, and Kalaba (1966) while the results for a spherical shell for a parallel input have been derived by Bellman, Kagiwada, Kalaba, and Ueno (1969).

11.4. THE SOLUTION OF THE SPHERICAL PROBLEM

As would be expected the most success in obtaining numerical results has been for the spherical shell problem with large values of the inner radius, r_0. Bellman, Kagiwada, and Kalaba (1966) show two methods for obtaining results.

The first procedure is to numerically evaluate the angular derivatives. This is done by utilizing the regular Gaussian integration positions for the range 0 to 1 and then calculating a numerical derivative matrix of coefficients for a L-1 degree polynomial. Thus*

$$\int_0^1 d\mu_i \, \hat{R}(r, \mu_i, \mu_j) = \sum_{i=1}^{L} w_i \, R_{ij}(r) \tag{11.4-1}$$

---

*In discrete form the ^-symbol has been dropped.

and

$$\frac{\partial}{\partial \mu_i} \hat{R}(r, \mu_i, \mu_j) = \sum_{l=1}^{L} a_{il} R_{lj}(r). \tag{11.4-2}$$

These coefficients, a_{il}, are tabulated in the above reference for several common Gaussian orders. Table 11 gives these coefficients for Gaussian and Radau forms of numerical integration positions in the range zero to one.

Equation (11.3-7) when transformed by Eq. (11.3-11) becomes*

$$\frac{d}{dr} R_{ij}(r) = -\frac{1-\mu_i^2}{\mu_i r} \sum_{l=1}^{L} a_{il} R_{lj}(r) - \frac{1-\mu_j^2}{\mu_j r} \sum_{l=1}^{L} a_{jl} R_{il}(r)$$

$$+ \frac{\mu_i^2 + \mu_j^2}{\mu_i^2 \mu_j^2 r} R_{ij}(r) - \left[\frac{1}{\mu_i} + \frac{1}{\mu_j}\right] R_{ij}(r)$$

$$+ c \left\{\hat{f}_{ij} + \sum_{l=1}^{L} w_l f_{lj} R_{il}(r)/\mu_l + \sum_{s=1}^{L} w_s R_{sj}(r) f_{is}/\mu_s \right.$$

$$+ \sum_{s=1}^{L} w_s R_{sj}(r)/\mu_s \sum_{l=1}^{L} w_l \hat{f}_{ls} R_{il}(r)/\mu_l\} \quad (i, j = 1, 2, \ldots, L).$$

$$\tag{11.4-3}$$

However, the above reference only utilizes isotropic scattering. The boundary conditions are

$$R_{ij}(r_0) = 0. \tag{11.4-4}$$

The referenced calculations utilize an r_0 value of 50 mean free paths so that the curvature is not large and this evaluation by numerical differentation appears satisfactory.

The second scheme utilized is a perturbation approach whereby an expansion in powers of r^{-n} is utilized. Thus let

$$\hat{R}(r, \mu, \mu_0) = \sum_{n=0}^{\infty} \frac{R^n(r, \mu, \mu_0)}{r^n}. \tag{11.4-5}$$

Here the R^0-term naturally corresponds to the results for slab geometry while the additional equations are of unpredicted form. For instance, for isotropic scattering the first two equations are in functional form

$$\frac{\partial}{\partial r} R^0(r, \mu, \mu_0) = -\left[\frac{1}{\mu} + \frac{1}{\mu_0}\right] R^0(r, \mu, \mu_0) + \frac{c}{2} \Gamma^0(\mu) \Gamma^0(\mu_0) \tag{11.4-6}$$

---

*The $\hat{}$ symbol on the f-term indicates when backscattering occurs.

TABLE 11

Numerical Differentiation Weights^a for Gaussian and Radau Integration Positions

Gaussian, $L = 2$
  μ_i = (0.2113248654405 19, 0.7886751 3459481)
  $a_{ij}{}^b$ = (-1.7320508075689, -1.7320508075689, 1.7320508075689, 1.7320508075689)

Gaussian, $L = 3$
  μ_i = (0.11270166537926, 0.50000000000000, 0.88729833462074)
  $a_{ij}{}^b$ = (-3.8729833462074, -1.2909944487358, 1.2909944487358,
    5.1639777949432, 0.0000000000000, -5.1639777949432,
    -1.2909944487358, 1.2909944487358, 3.8729833462074)

Gaussian, $L = 5$
  μ_i = (0.0469100770, 0.2307653449, 0.5000000000, 0.7692346551, 0.9530899230)
  $a_{ij}{}^b$ = (-10.134081191, -1.920512047, 0.6023336319, -0.488833230, 1.103533701,
    15.403904170, -1.516707434, -2.8707764846, 1.857116053, -3.920798231,
    -8.087087509, 4.805501304, 0.0000000000, -4.805501304, 8.087087509,
    3.920798231, -1.857116053, 2.8707764846, 1.516706434, -15.403904170,
    -1.103533701, 0.488833230, -0.602336319, 1.902512047, 10.134081191)

Radau, $L = 2$
  μ_i = (0.3333333333333333, 1.000000000000000)
  $a_{ij}{}^b$ = (-1.5000000000000000, -1.5000000000000000, 1.5000000000000000,
    1.5000000000000000)

Radau, $L = 3$
  μ_i = (0.15505102572168, 0.64494897427832, 1.00000000000000)
  $a_{ij}{}^b$ = (-3.22474487139160, -0.85773803324704, 0.85773803324704,
    4.85773803324704, -0.77525512860840, -4.85773803324704,
    -1.63299316185544, 1.63299316185544, 4.00000000000000)

Radau, $L = 5$
  μ_i = (0.057104196114, 0.276843013638, 0.583590432369, 0.860240135656,
    1.000000000000)
  $a_{ij}{}^b$ = (-8.75592397794, -1.47717250914, 0.40335296739, -0.25747222895,
    0.48038472266, 14.02023254656, -1.80607772408, -2.13281586098,
    1.09198625336, -1.92966670916, -8.94418347712, 4.98293020975,
    -0.85676524542, -3.51979172716, 5.12223987336, 6.02131692059,
    -2.69063175880, 3.71212520772, -0.58123305256, -15.67295788686,
    -2.34144201209, 0.99095178227, -1.12589706871, 3.26651075531,
    12.0000000000)

^{a}The reported digits of accuracy are believed to be correct from calculations in double precision, 16 digits, on an IBM 360/50 computer, thus the accuracy varies depending upon the order of the calculation.
^{b}The FORTRAN method of stepping i first and j second is employed in listing the results as a single vector, i.e., the order is a_{11}, a_{21}, ..., a_{n1}, a_{12}, a_{22}, ..., a_{n2}, ..., a_{1n}, a_{2n}, ..., a_{nn}.

and

$$\frac{\partial}{\partial r} R^1(r, \mu, \mu_0) = - \left[\frac{1}{\mu} + \frac{1}{\mu_0}\right] R^1(r, \mu, \mu_0)$$

$$- \frac{1 - \mu^2}{\mu} \frac{\partial}{\partial \mu} R^0(r, \mu, \mu_0) - \frac{1 - \mu_0^2}{\mu_0} \frac{\partial}{\partial \mu_0} R^0(r, \mu, \mu_0)$$

$$+ \frac{\mu^2 + \mu_0^2}{\mu^2 \mu_0^2} R^0(r, \mu, \mu_0) + \frac{c}{2} \Gamma^0(\mu) \int_0^1 d\mu'' R^1(r, \mu'', \mu_0)/\mu''$$

$$+ \frac{c}{2} \Gamma^0(\mu_0) \int_0^1 d\mu' R^1(r, \mu, \mu')/\mu', \qquad (11.4\text{-}7)$$

where

$$\Gamma^0(\mu) = 1 + \int_0^1 d\mu' R^0(r, \mu, \mu')/\mu'$$

$$= 1 + \int_0^1 d\mu'' R^0(r, \mu'', \mu)/\mu''. \qquad (11.4\text{-}8)$$

Conceptually the R^0-equation can be solved first and the result utilized in the R^1-equation, etc. Note that the partial derivatives with respect to the angles are always on the previous solution; thus numerically the values can be smoothed if necessary in order to retain stability of the numerical system. The boundary condition is naturally

$$R^n(r_0, \mu, \mu_0) = 0. \qquad (11.4\text{-}9)$$

Again both of these procedures work if r_0 is kept large.

Another numerical approach to the solution of Eq. (11.3-7) for the further restrictions to isotropic scattering has been utilized by Allen, Shampine and Wing (1970). Here four different finite difference algorithms are proposed along with appropriate step-ahead schemes for their solution. Thus for the form

$$\left[\frac{\partial}{\partial r} + \frac{1 - \mu^2}{\mu r} \frac{\partial}{\partial \mu} + \frac{1 - \mu_0^2}{\mu_0 r} \frac{\partial}{\partial \mu_0} - \frac{\mu^2 + \mu_0^2}{\mu^2 \mu_0^2 r} + \frac{1}{\mu} + \frac{1}{\mu_0}\right] \hat{R}(r, \mu, \mu_0) = \frac{c}{2} \Gamma(\mu) \Gamma(\mu_0),$$

$$(11.4\text{-}10)$$

where

$$\Gamma(\mu) = 1 + \int_0^1 d\mu' \hat{R}(r, \mu, \mu')/\mu' = 1 + \int_0^1 d\mu'' \hat{R}(r, \mu'', \mu)/\mu'', \qquad (11.4\text{-}11)$$

the finite difference result is

$$n\Delta r\, ijh^2 \left[\frac{R_{ij}^n - R_{ij}^{n-1}}{\Delta r}\right] + jh(1 - i^2 h^2) \left[\frac{R_{ij}^n - R_{i-1,j}^n}{h}\right]$$

$$+ (ih)(1 - j^2 h^2) \left[\frac{R_{ij}^n - R_{i,j-1}^n}{h}\right] + n\Delta r\, h(i + j) R_{ij}^n$$

$$= n\Delta r \, ijh^2 \, \frac{c}{2} \, [1 + \sum_k w_k \, R_{ik}^{n-1}/(kh)] \; [1 + \sum_k w_k \, R_{kj}^{n-1}/(kh)]$$

$$+ [h^2(i^2 + j^2)/(ijh^2)] \, R_{ij}^{n-1} \quad \left(n = 1, 2, \ldots ; i,j = 1, 2, \ldots, \frac{1}{h} \right).$$

$$(11.4\text{-}12)$$

Here $h = \Delta\mu = \Delta\mu_0$ and Eq. (11.4-10) is multiplied by $r\mu\mu_0$ for the above form. The difference nomenclature is

$$R_{ij}^n = R(n\Delta r, ih, jh) = \hat{R}(r, \mu, \mu_0). \qquad (11.4\text{-}13)$$

Because of the finite differences used for the angular variables, the numerical integration formula utilized must be for equal spacing such as the trapezoidal rule or Simpson's rule. By using the zero-end-point these procedures imply that

$$\lim_{\mu \to 0} \frac{R(r, \mu, \mu_0)}{\mu} = 0 \qquad (11.4\text{-}14)$$

with a like expression for μ_0.

Equation (11.4-12) can then be solved for R_{ij}^n in terms of R_{ij}^{n-1} and thus a step ahead procedure is generated. The initial condition is reasoned as being the surface of a small sphere of radius ϵ whose material is totally absorbing and whose cross section is infinite. Thus the zero initial condition is

$$R_{ij}^0 = R_{0j}^n = R_{i0}^n = 0. \qquad (11.4\text{-}15)$$

This boundary condition appears to contradict Eq. (11.3-9) and thus the center condition for a sphere has not in general been resolved.

Allen, Shampine, and Wing (1970) also give three other finite difference forms that could be utilized. From their work it appears that Δr and h need to be quite small as with $h = 0.02$ and $\Delta r = 0.00125$, an accuracy of three places is reported for the albedo.

Because of the conservative nature of Eq. (11.3-12) it would appear that a finite difference approach to this equation would be feasible; however, this has not been reported at the present time.

A limited number of numerical evaluations for spherical shells have been carried out by Bellman, Kagiwada and Kalaba (1966) and by Allen, Shampine and Wing (1970) but extensive listings of general results for spherical geometry are not available. In addition, a general all purpose numerical procedure does not appear to yet exist for this curved geometry case.

11.5. EXTENSIONS TO OTHER CURVED GEOMETRIES

Because the results for the simplest curved geometry, the symmetric sphere, have been less than desirable, the extension to other curved geometries has been sparse. Bellman, Kagiwada, Kalaba, and Ueno (1969) have extended the analysis to a nonsymmetric spherical problem caused not by varying medium properties, but by a parallel beam input; however they report no numerical results. Kalaba and Ruspini (1969) deduce the applicable integral equation for a homogeneous cylinder and attack it by an initial value method. Thus an invariant imbedding form is obtained; however the particle counting procedure is not attempted here.

In summary the application of invariant imbedding theory to the transport problems of curved geometries remains a "wide open" field.

11.6. EXERCISES

1. Formulate the invariant imbedding reflection function problem for the case of a spherical shell starting at radius, r_0, where the inner sphere is a vacuum.

2. In reference to Eqs. (11.4-6, 7) deduce the equation for the $R^2(r, \mu, \mu_0)$ functional and for the general $R^n(r, \mu, \mu_0)$ term.

3. Derive the albedo equation for a sphere with a parallel beam of particles as the input by defining a central angle, ψ, such that $\psi = 0$ is directly parallel to the incoming beam.

4. Derive the invariant imbedding reflection function equation for a symmetric cylinder of infinite height and compare with the form of Bellman, R. E., et al. (1960). Deduce any omitted terms.

5. Compare the albedo for a sphere given by Bellman, R. E., et al. (1960) with that given in this chapter and note the difference as explained by Bailey and Wing (1964A).

CHAPTER 12

APPENDIX: NUMERICAL ANALYSIS CONSIDERATIONS

12.1. ORDINARY DIFFERENTIAL EQUATIONS

Since the discrete form of many of the invariant imbedding transport equations utilized in this book results in nonlinear ordinary differential equations, the numerical procedures that have been successful in past calculations are discussed briefly. Naturally there are many fine treatises on this subject such as the work of Collatz (1966); however, an extensive bibliography on this subject will not be attempted here.

In general only Runge-Kutta methods for the numerical solution of ordinary differential equations are discussed as they apply to the problems posed by this book. To understand this it is necessary to note that the class of ordinary differential equations to which this particular invariant imbedding set belongs is commonly referred to as a "stiff" system of differential equations. This connotation arose from the work of Curtis and Hirschfelder (1952) in working with systems of chemical kinetic equations. In essence, this name implies that the characteristic responses of the system will differ by many orders of magnitude. If the system of equations is mathematically linear, then these responses represent the characteristic values of the solution; however, for nonlinear systems no direct analogy exists except to indicate a similar overall response. When a numerical solution is attempted for systems of stiff differential equations, it becomes necessary to utilize a small integration step geared to the most rapidly varying system response and many times the step size is prohibitively small. Thus, much computer "power" is required when a large system, i.e., several thousand simultaneous equations, is solved for any reasonable thickness of material.

One property of Runge-Kutta methods is that they are relatively "stable" in relationship to this stiff differential equation problem, particularly when compared to certain predictor-corrector methods, for instance see Hamming (1962). In addition Runge-Kutta methods require a minimum of storage of arrays as well as a simple program for execution. In practice these Runge-Kutta procedures have proven to be very useful in obtaining numerical values for the invariant imbedding equations.

Some estimate as to how "stiff" these invariant imbedding equations are can be gained by reference to the quantity

$$\tau_{ij} = 1/\mu_i + 1/\mu_j \qquad (12.1\text{-}1)$$

TABLE 12

τ-value of Various Order Radau
Integration Positions

L	$(\tau_{ij})_{\max}$	$(\tau_{ij})_{\min}$
2	6.0	2.0
3	12.9	2.0
4	22.6	2.0
5	35.0	2.0

which is Eq. (8.7-32). For a Radau set of τ-values, see Table 1; the maximum and minimum values are shown in Table 12. The situation is worse when the energy dependent case is considered as group total cross sections will appear in the expression, for instance see Eq. (5.2-8). In addition it is noted that as the order of the system increases, i.e., L becomes larger, the system also becomes more stiff in nature.

This τ-parameter is the exponential response to the single scattered albedo result as is shown by Eq. (5.3-11) and represents a good measure of the stiffness of any invariant imbedding albedo equation.

12.2. RUNGE-KUTTA-GILL METHOD

The most popular of the Runge-Kutta methods is probably the Runge-Kutta-Gill procedure which has been specifically designed in order to minimize the storage requirements; therefore, it is especially useful for large systems of ordinary differential equations. This procedure is of fourth-order in accuracy so that a highly accurate computed result can be expected. A good reference to this procedure is that by Ralston and Wilf (1960) where they show the complete derivation of this method as well as some other Runge-Kutta methods. In addition, a subroutine flow chart for this procedure is shown. The Runge-Kutta-Gill method as utilized is

$$y' = f(x, y), \qquad y_0 = y(x_0)$$

$$k_1 = hf(x_0, y_0)$$

$$y_1 = y_0 + \frac{1}{2}(k_1 - 2q_0)$$

$$q_1 = q_0 + 3\left[\frac{1}{2}(k_1 - 2q_0)\right] - \frac{1}{2}k_1$$

$$k_2 = hf(x_0 + \frac{1}{2}h, y_1)$$

$$y_2 = y_1 + (1 - \sqrt{1/2})(k_2 - q_1)$$
$$q_2 = q_1 + 3[(1 - \sqrt{1/2})(k_2 - q_1)] - (1 - \sqrt{1/2})k_2$$
$$k_3 = hf(x_0 + \tfrac{1}{2}h, y_2)$$
$$y_3 = y_2 + (1 + \sqrt{1/2})(k_3 - q_2) + \tfrac{1}{6}(k_4 - 2q_3)$$
$$q_3 = q_2 + 3[(1 + \sqrt{1/2})(k_3 - q_2)] - (1 + \sqrt{1/2})k_3$$
$$k_4 = hf(x_0 + h, y_3)$$
$$y_4 = y_3 + \tfrac{1}{6}(k_4 - 2q_3)$$
$$q_4 = q_3 + 3\left[\tfrac{1}{6}(k_4 - 2q_3)\right] - \tfrac{1}{2}k_4. \tag{12.2-1}$$

Initially q_0 is zero and q_4 replaces q_0 for each subsequent step. This procedure attempts to compensate for round-off error as q_4 would be identically zero if infinite wordlength calculations are performed. The answer for $y(x_0 + h)$ is naturally y_4. In the generalization to a system of equations the various quantities are written in vector form; therefore if the system is of order N, the storage requirements are 3N.

The error estimate for the Runge–Kutta–Gill method is not amenable to easily calculational procedures. The most convenient procedure is an evaluation with two h-sizes and then to apply a Richardson extrapolation procedure as shown by Moursund and Duris (1967) to produce

$$e_n = \frac{y_n(h/2) - y_n(h)}{30}, \tag{12.2-2}$$

where the error is for one integration step. The notation means utilizing the indicated h-size for determining the same y-value. This procedure naturally consumes considerable computer time and thus is not readily employed.

12.3. RUNGE–KUTTA–MERSON METHOD

Another useful variant to the general Runge–Kutta method is the modification by Merson that is designed to easily allow an error evaluation. This procedure is summarized in one form by Fox (1962) and in another form by Chai (1970). The latter version appears more usable since it allows the determination of an adequate h-size to maintain the required accuracy. This is summarized as

$$y' = f(x, y), \quad y_0 = y(x_0)$$
$$k_1 = h/3\, f(x_0, y_0)$$
$$k_2 = h/3\, f(x_0 + h/3, y_0 + k_1)$$

$$k_3 = h/3 \, f(x_0 + h/3, y_0 + 1/2 \, k_1 + 1/2 \, k_2)$$
$$k_4 = h/3 \, f(x_0 + h/2, y_0 + 3/8 \, k_1 + 9/8 \, k_3)$$
$$k_5 = h/3 \, f(x_0 + h, y_0 + 3/2 \, k_1 - 9/2 \, k_3 + 6 \, k_4)$$
$$y_1^{(p)} = y_0 + 3/2 \, k_1 - 9/2 \, k_3 + 6 \, k_4$$
$$e_1 = k_1 - 9/2 \, k_3 + 4 \, k_4 - 1/2 \, k_5$$
$$e_1 > e_{max}; \quad h = h/2; \quad \text{repeat.}$$
$$e_{max} > e_1 > e_{max}/16; \quad y_1 = y_1^{(p)} - e_1; \quad h = h; \quad \text{continue}$$
$$e < e_{max}/16; \quad y_1 = y_1^{(p)} - e_1; \quad h = 2h; \quad \text{continue} \qquad (12.3\text{-}1)$$

This sequence allows the automatic control of the step size where the maximum error per step, e_{max}, is realistically chosen. It is noted though that the total error, E, for any calculation cannot be estimated from this per step value without further information. The quantity

$$E = S \, e_{max}, \qquad (12.3\text{-}2)$$

where S is the total number of integration steps, usually higher over-estimates the error. Again for a system of equations the appropriate quantities become vectors.

12.4. EXAMPLE CALCULATIONS

In applying the Runge–Kutta methods to the numerical solution of the discrete forms of the invariant imbedding equations, the recommended procedure is to do a typical calculation using both the Runge–Kutta–Gill and Runge–Kutta–Merson algorithms. Naturally the h-values must be within the stability range of the general fourth-order Runge–Kutta methods. From the Merson version the per step error is controlled and the apparent total error for different e_{max} values can be deduced from the solution values at corresponding x-values. Once an apparent satisfactory h-value is deduced, the Gill version is run and the final solution values compared. When these are acceptable the Gill version is utilized to perform the remainder of the calculations. The following example calculations for the albedo indicate how this procedure is applied.

The numerical solution of Eq. (5.2-10) is performed with the script-$\mathfrak{R}$ transformation of Eqs. (5.3-13, 14). The restriction to isotropic scattering is imposed as well as limiting the calculation to two energy groups, $N = 2$. In addition a Gaussian order scheme of three, $L = 3$, is employed as per Table 1. Thus the system is of order 36. The total cross sections are $\Sigma_1 = 0.05$ and $\Sigma_2 = 5.0$, both in cm^{-1} units, while the c-matrix is

$$c = \begin{pmatrix} 0.3 & 0.3 \\ 0.5 & 0.4 \end{pmatrix} \qquad (12.4\text{-}1)$$

This combination of total cross section values and Gaussian order makes the τ-values

$$\tau_{min} = 0.113 \qquad (12.4\text{-}2)$$

and

$$\tau_{max} = 88.7, \qquad (12.4\text{-}3)$$

and thus the system is quite stiff in nature.

The calculations are performed over the range $0 \leqslant x \leqslant 1.0$ cm as an example. The results for various x-values near unity are compared for $\mathcal{R}$-values as shown by Table 13. It is noted that since the Merson version has a variable step size that the printed value may not correspond exactly to the desired value; however, good insight can be gained from the computed results in any manner.

TABLE 13

Selected Albedo Results for Runge–Kutta Integration

Type	h_{min}	h_{max}	e_{max}	x-value	$\mathcal{R}_{11}^{11}$	$\mathcal{R}_{12}^{11}$	$\mathcal{R}_{21}^{11}$	$\mathcal{R}_{22}^{11}$	Time[a]
RKM	0.02	0.02	–	1.000	0.008850	0.023384	0.000389	0.012663	1.26
RKM	0.005	0.08	0.0005	1.005	0.007191	0.023195	0.000386	0.012664	0.86
RKM	0.0025	0.08	0.0001	0.985	0.007075	0.023180	0.000386	0.012663	0.99
RKM	0.0025	0.04	0.00005	1.005	0.008588	0.023356	0.000389	0.012663	1.17
RKM	0.00125	0.02	0.00001	0.994	0.008394	0.023335	0.000388	0.012663	1.76
RKG	0.02	0.02	–	1.000	0.008850	0.023384	0.000389	0.012663	1.00
RKG[b]	0.04	0.04	–	1.000	0.008805	0.070822	0.001180	−0.827376	0.61

[a]Relative time compared to standard RKG, $h = 0.02$.
[b]Unstable case.

Table 13 vividly shows the difficulty in solving stiff systems in that different dependent variables are computed to different accuracies. From these results it is apparent that $\mathcal{R}_{11}^{ij}$ are the governing values near $x = 1.0$; however, at small x-values the $\mathcal{R}_{22}^{i}$ will be controlling. The RKG method with $h = 0.02$ appears to give satisfactory and efficient results for this particular case.

It is noted that in the RKM method with variable step size that the input h-value is 0.02 and that the value is halved down to h_{min} for the first step and then doubling occurs with repeated steps. The value of e_{max} required for good accuracy has to be found by trial and error although previous experience with similar calculations can be a big help.

The calculations shown in this book are performed with the RKG routine with $h = 0.02$ normally while some selected values are obtained with $h = 0.01$.

REFERENCES

Abramowitz, M., et. al. (1964), "Handbook of Mathematical Functions," National Bureau of Standards, AMS 55, U.S. Government Printing Office, Washington, D.C.

Ambarzumian, V. A. (1943), Diffuse reflection of light by a foggy medium, *Compt. Rend. Acad. Sci. SSSR* **38**, 229.

Aronson, R., and Yarmush, D. L. (1966), Transfer matrix method for gamma ray and neutron penetration, *J. Math. Phys.* **7**, 221.

Bailey, P. B., and Wing, G. M. (1964), A rigorous derivation of some invariant imbedding equations of transport theory, *J. Math. Anal. Appl.* **8**, 144.

Bailey, P. B., and Wing, G. M. (1964A), A correction to some invariant imbedding equations of transport theory obtained by 'Particle Counting', *J. Math. Anal. Appl.* **8**, 170.

Bailey, P. B., and Wing, G. M. (1965), Some recent developments in invariant imbedding with applications, *J. Math. Phys.* **6**, 453.

Bellman, R. E. (1957), "Dynamic Programing," Princeton University Press, Princeton, N.J.

Bellman, R. E. (1967), "Introduction to the Mathematical Theory of Control Processes," Vol. 1, Academic Press, New York.

Bellman, R. E. (1967A), Invariant imbedding and computational methods in radiative transfer, University of Southern California, Electronic Sciences Laboratory Report USCEE-218.

Bellman, R. E., et. al. (1960), Invariant imbedding and mathematical physics: I. Particle processes, *J. Math. Phys.* **1**, 280.

Bellman, R. E., and Kalaba, R. E. (1956), On the principle of invariant imbedding and propagation through inhomogeneous media, *Proc. Nat. Acad. Sci. USA* **42**, 629.

Bellman, R. E., Kalaba, R. E., and Prestrud, M. C. (1963), "Invariant Imbedding and Radiative Transfer in Slabs of Finite Thickness," American Elsevier, New York.

Bellman, R. E., Kagiwada, H. H., and Kalaba, R. E. (1966), Invariant imbedding and radiative transfer in spherical shells, *J. Comput. Phys.* **1**, 245.

Bellman, R. E., Kagiwada, H. H., Kalaba, R. E., and Ueno, S. (1969), Diffuse reflection of solar rays by a spherical shell atmosphere, *Icarus* **11**, 417.

Carlson, B., and Bell, G. I. (1958), Solution of the transport equation by the S_N Method, *Proc. Intern. Conf. Peaceful Uses of At. Energy Geneva* **16**, 535.

Carlvik, I. (1967), Monoenergetic critical parameters and decay constants for small spheres and thin slabs, Aktiebologet Atomoenergic, Report AE-273, Stockholm, Sweden.

Chai, A. S. (1970), Comment on Runge-Kutta-Merson algorithm, *Simulation* **15**, 89.

Chandrasekhar, S. (1950), "Radiative Transfer," Oxford University Press, New York.

Collatz, L. (1966), "The Numerical Treatment of Differential Equations," 3rd ed., Springer-Verlag, New York.

Curtis, C. F., and Hirschfelder, J. O. (1952), Integration of stiff equations, *Proc. Natl. Acad. Sci. USA* **38**, 235.

Davison, B. (1958), "Neutron Transport Theory," Oxford University Press, New York.

Denman, E. D. (1970), "Coupled Modes in Plasmas, Elastic Media and Parametric Amplifiers," American Elsevier, New York.

Dodson, D. S., and Mingle, J. O. (1964), Escape probabilities by the method of invariant imbedding, *Trans. Amer. Nucl. Soc.* **7**, 26.

Fano, U., Spencer, L. V., and Berger, M. J. (1959), Penetration and Diffusion of X Rays, "Handbuch der Physik," Vol. 38/2, 660, Springer-Verlag, New York.

/

Fox, L. (1962), "Numerical Solution of Ordinary and Partial Differential Equations," Pergamon Press, Oxford.

Francis, N. S., et. al. (1958), Variational solutions of the transport equation, *Proc. Intern. Conf. Peaceful Uses of At. Energy Geneva* **16**, 517.

Froberg, C. E. (1965), "Introduction to Numerical Analysis," Addison-Wesley, Reading, Mass.

Glasstone, S., and Sesonke, A. (1963), "Nuclear Reactor Engineering," Van Nostrand, Princeton, N.J.

Hamming, R. W. (1962), "Numerical Methods for Scientists and Engineers," McGraw-Hill, New York.

Honeck, H. C. (1967), ENDF/B specifications for an evaluated nuclear data file for reactor applications, rev., AEC Report BNL-50066.

Kaiser, R. E. (1968), Invariant Imbedding Theory of Photon Transport, Ph.D. Dissertation, Kansas State Univ., Manhattan, Ks.

Kaiser, R. E., and Mingle, J. O. (1967), Reflection of high-energy photons from semi-infinite homogeneous slabs by invariant imbedding techniques," *Trans. Amer. Nucl. Soc.* **10**, 733.

Kalaba, R. E., and Ruspini, E. (1969), Invariant imbedding and radiative transfer in a homogeneous cylindrical medium, University of Southern California, Electronic Sciences Laboratory Report USCEE-349.

Kaplan, I. (1962), "Nuclear Physics," 2nd. ed., Addison-Wesley, Reading, Mass.

Mingle, J. O. (1966), Reflection and transmission intensities from gray slabs utilizing invariant imbedding methods, *C.E.P. Symposium Series* **62**, No. 64, 250.

Mingle, J. O. (1967), Application of the invariant imbedding method to monoenergetic neutron transport theory in slab geometry, *Nucl. Sci. Eng.* **28**, 177.

Mingle, J. O. (1969), Introduction to the invariant imbedding method for curved geometries, Neutron Transport Theory Conference, Virginia Polytechnic Institute, Blacksburg, Va., AEC Report ORO-3858-1, 640.

Mingle, J. O. (1970), Approximate solutions to nonlinear integral equations, *J. Math. Anal. Appl.* **32**, 651.

Moursund, D. G., and Duris, C. S. (1967), "Elementary Theory and Application of Numerical Analysis," McGraw-Hill, New York.

Ralston, A., and Wilf, H. S. (1960), "Mathematical Methods for Digital Computers," Vol. 1, John Wiley, New York.

Rosescu, T. (1966), Invariant imbedding in neutron transport theory, Academia Republicii Socialiste Romania, Institutul de Fizica Atomica, FR-57.

Shimizu, A. (1967), Tabulation of dose transmission factors for slabs, NBS Reports 9617, 9618.

Shimizu, A., and Mizuta, H. (1966), Application of invariant imbedding to the reflection and transmission problem of gamma rays, *J. Nucl. Sci. Techn. (Japan)* **3**, 57.

Storm, E. and Israel, H. I. (1967), Photon cross sections from 0.001 to 100 Mev for elements 1 through 100, AEC Report LA-3753.

Timmons, D. H., and Mingle, J. O. (1967), The application of invariant imbedding to Green's function, *Trans. Amer. Nucl. Soc.* **9**, 437.

Warming, R. F., and Heaslet, M. A. (1966), Radiation flux from a slab or sphere, *J. Math. Anal. Appl.* **14**, 359.

Weinberg, A., and Wigner, E. P. (1958), "The Physical Theory of Neutron Chain Reactors," Univ. of Chicago Press, Chicago, Ill.

Wing, G. M. (1961), Invariant imbedding and transport theory—A unified approach, *J. Math. Anal. Appl.* **2**, 277.

Wing, G. M. (1962), "An Introduction to Transport Theory," John Wiley, New York.

INDEX

JOHN ORVILLE MINGLE is currently the director of the Institute for Computational Research in Engineering and a professor of nuclear engineering at Kansas State University. In addition to this teaching experience, Dr. Mingle has been a consultant for several industrial companies.

Dr. Mingle received his B.S. and M.S. degrees from Kansas State University and his Ph.D. from Northwestern University. He is an active member of the American Nuclear Society, the Society for Industrial and Applied Mathematics, the National Society of Professional Engineers, the American Institute of Chemical Engineers, the Society of Sigma XI, and the American Society for Engineering Education. He is a regular contributor to a number of technical publications.